INSTALLER WORKBOOK

For ACI Certification Program:
Adhesive Anchor Installer

Reported by ACI Certification Programs Committee

Joe Hug, Chair

Bryan R. Castles	Augusto H. Holmberg	Christopher J. Robinson
William Ciggelakis	J. Scott Keim	Xiomara Sapon
Oscar Duckworth	Steve R. Lloyd Sr.	Robert L. Varner
Werner K. Hellmer		Wayne M. Wilson

John W. Nehasil, Staff Liaison

Developed by ACI Committee C680—Adhesive Anchor Installer

W. Calvin McCall, Chair

Ronald L. Kozikowski Jr., Secretary

Sean L. Agresta	Werner A. F. Fuchs	Stephen Marchese
Neal S. Anderson	Werner K. Hellmer	Donald F. Meinheit
TJ Bland	Andra Hoermann-Gast	George W. Seegebrecht
Jay Dorst	Scott B. Hougard	John F. Silva
Robert J. Evans	Amy S. Kolczak	Nathan Westin

Michael J. Morrison, Staff Liaison

Associate Members

Jacques A. Bertrand	Pedro N. Quiroga	Siobhan Sawyer
John E. Pearson	Thomas L. Rozsits	

Consulting Members

Gregory S. Christensen	Miguel C. Mota	Alexander P. Wehbe
Christopher La Vine	Jake Olsen	

PUBLICATION CP-80, 2ND EDITION
AMERICAN CONCRETE INSTITUTE
FARMINGTON HILLS, MICHIGAN

i

The ACI Certification Programs Committee wishes to express its thanks to the ACI Educational Activities Committee, ACI Committee E 905 – Training Programs, and the Task Group of the Subject Matter Experts coordinated by Thomas O. Malerk who made the expedited production of this Workbook possible.

Editor
Claire A. Freeman

Second Edition
Seventh Printing, January 2024
Copyright © 2020
American Concrete Institute
38800 Country Club Drive
Farmington Hills, Michigan 48331
Phone (248) 848-3700 • FAX (248) 848-3701

PREFACE

Adhesive Anchors are recognized by the structural design profession as an important structural connection in many applications, with design and installation interdependent in achieving proper performance. The American Concrete Institute's 2011 Building Code Requirements for Structural Concrete, which is referenced by the 2012 International Building Code, recognizes the role installation plays in proper performance of anchors and now requires certification of installers when installing adhesive anchors in specific orientations. To respond to this need for certification, the American Concrete Institute developed and maintains a program to certify Adhesive Anchor Installers.

The ACI Certification Program for Adhesive Anchor Installer identifies Installers who have demonstrated the ability to read, comprehend, and execute instructions to properly install adhesive anchors in concrete. The Adhesive Anchor Installer must also possess the knowledge to properly assess ambient conditions, the condition of the concrete, materials, equipment, and tools for installing adhesive anchors and determine when it is appropriate to proceed with installation or when additional guidance from a supervisor/foreman/ project engineer is needed.

Although written for Installers, much of the information in this Workbook will be of value for Concrete Construction Special Inspectors as continuous inspection of adhesive anchor installation grows in frequency, and may also subject to future credentials.

This Installer Workbook (CP-80) describes the operation of the ACI Certification Program and provides the resource materials and directions needed to prepare for the examinations.

Check the ACI website for updated information and/or errata/addendums:
https://www.concrete.org/publications/documenterrata.aspx

CONTENTS

PROGRAM INFORMATION

Purpose and Importance of ACI Certification

Adhesive anchor effectiveness is measured by the bond strength achieved between the adhesive and concrete, and between the adhesive and anchor. Adhesive anchor manufacturers have developed installation procedures for their specific products, that when followed, are intended to provide the proper conditions for the anchor system to achieve that required bond strength.

The ACI Adhesive Anchor Installer Certification program is designed to verify that certified individuals possess the necessary skills to properly install adhesive anchors in concrete.

Certification as an Adhesive Anchor Installer through ACI provides you with international recognition by an independent group of acknowledged concrete experts. This recognition may improve your potential for advancement and may provide you with more employment opportunities.

Overview of the ACI Program

To attain ACI Certification as an Adhesive Anchor Installer, you must successfully complete both the ACI written and performance examinations (see Figure 1). While there are no prerequisites in either training or experience, the examinations are designed so that only knowledgeable and skilled installers will pass.

Administration

The American Concrete Institute, through its Certification Programs Committee, administers the certification program. This includes the development and maintenance of all program policies and procedures, instructional materials, and examinations.

The ACI certification program is conducted by a sponsoring group, such as an ACI chapter, a state or local concrete industry association, a college or university, or other organization committed to upgrading the quality of concrete. This group is responsible for scheduling and conducting all training courses and examination sessions; arrangement of facilities, material, equipment, and personnel; payment of all bills; and establishment and collection of all registration fees.

The ACI Certification Department assists the sponsoring groups in setting up the program and manages the daily operation of the program. This includes the publication and dissemination of all instructional materials and examinations, grading examinations, maintenance of certification records, and issuance of certification to successful candidates.

Definitions

ACI — The American Concrete Institute (38800 Country Club Drive, Farmington Hills, MI 48331), which developed and published instructional materials and examinations used in this program, and which certifies individuals who satisfy all requirements of this program.

ACI Adhesive Anchor Installer — an individual who has demonstrated the ability to read, comprehend, and execute instructions to properly install adhesive anchors in concrete.

ACI Certification — A formal recognition, valid for a specified amount of time, which shows that a person has satisfactorily completed the certification requirements.

ACI Certification Department — The department within ACI that manages the certification programs.

Examiner — An individual approved by ACI and authorized to administer the examination sessions locally.

Recertification — A process, undertaken every five years after original certification, by which a certified individual renews certification by successfully completing the then-current recertification requirements.

Sponsoring Group — The local organization (such as an ACI chapter, trade association, public agency, private testing laboratory, or other organization) that takes responsibility for conducting the certification examination in its locality and, as a recommended option, for conducting a training course.

Supplemental Examiner — A person who has been deemed qualified by the examiner who directly observes and evaluates the examinee's performance during the performance examination.

Trial — In the performance examination, one of two attempts allowed for the examinee to successfully perform each step of the procedure.

Educational Materials

The Adhesive Anchor Installer Workbook contains all the materials you will need to prepare for the examination. This Program Information section of the Installer Workbook describes the certification program and its policies and procedures.

ACI ADHESIVE ANCHOR
INSTALLER CERTIFICATION
PROCESS

Figure 1

DECIDE TO PURSUE CERTIFICATION & LOCATE SPONSORING GROUP

TAKE OPTIONAL REVIEW COURSE OR SELF-GUIDED STUDY

PASSED WRITTEN EXAM?

YES

NO

PASSED PERFORMANCE EXAM WITHIN ONE YEAR?

YES

NO

CERTIFICATION GRANTED

RECERTIFICATION
After Five Years

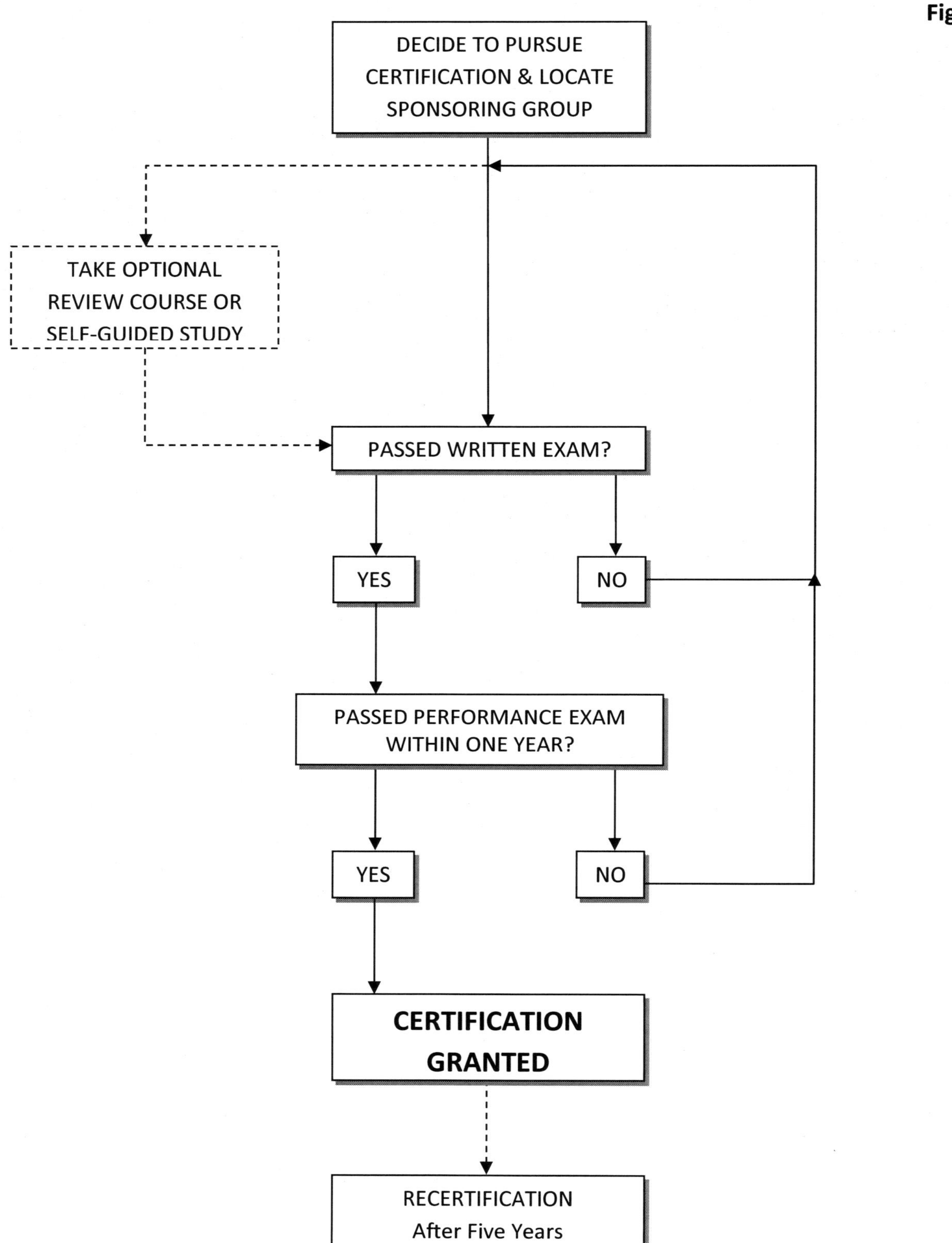

Job Task Analysis (JTA)

The JTA is a detailed list of fundamental job elements, including the items of knowledge and skills required to successfully perform a job's functions. These elements are further distinguished in the JTA between general knowledge (understanding concepts), working knowledge (knowing facts or values), and physical proficiency (performing a physical task).

The written and performance examinations for Adhesive Anchor Installer are designed to assess these fundamental job elements. Therefore, the JTA may also serve as a guide when studying/preparing for certification; it outlines the expected knowledge and skill base of an Adhesive Anchor Installer and governs the areas of competency assessed by the written and performance examinations for this program. The JTA is provided in the Job Task Analysis section of this reference.

Examinations

To achieve Adhesive Anchor Installer certification, both a written and performance exam must be passed. You may be required to show photo identification to gain admittance to the exams.

ACI uses an optical reader to scan all personal data, examination answer sheets, and program evaluation forms. Therefore, bring two No. 2 pencils and a clean eraser to the examination sessions.

These exams may be conducted verbally if requested; please make arrangements with the sponsoring group prior to the examination date.

Written Exam

The written exam covers all the materials listed in **Education Materials**. While the level of difficulty for each question will vary, they are not meant to trick or confuse you.

Calculations may be required for some questions; therefore, you may bring a simple-function, pocket calculator to the exam. You may not share a calculator with another examinee during the exam.

The exam is **closed book**, which means that no technical materials or notes are allowed in the room during the examination.

The exam consists of approximately 75 multiple-choice questions. A maximum of 1.5 hours (90 minutes) is allowed to complete the entire exam. A score of 74% or higher is required to pass the written examination.

Challenges

If you encounter a test question that appears to be unclear, incorrect, or unfair, please describe your complaint on the Question Challenge Form included with your examination.

You must mark a response on your answer sheet that best answers the question. All questions have been formulated from the Educational Materials; any challenge based on a reference from an outside document will not be considered unless the reference agrees with the Educational Materials If your explanation is found to be valid, scores will be adjusted accordingly.

All questions for the written exam are developed by the Adhesive Anchor Installer Certification Committee and maintained at ACI Headquarters by the ACI Certification Department. Neither the examiner nor the sponsoring group have any jurisdiction over the content or the scoring of the written exam.

Performance Exam

The Adhesive Anchor Installer performance exam covers vertical down and overhead piston plug systems. You will be judged on your ability to correctly perform all required steps for each procedure based on the criteria in the Performance Examination Checklists. In instances where performance is not feasible, you will be prompted to verbally describe the step(s). Checklists are provided in **Study Materials** to aid your study, but are revised periodically and may not be inclusive of the current exam.

Omission or incorrect performance on one or more steps of a procedure will constitute failure of that trial. You are allowed up to two trials on the day of the exam for each procedure. Failure of any procedure after two trials will constitute failure of the performance exam. All procedures must be passed at the same time to pass the performance examination.

You are required to perform the correct steps for each procedure in the direct presence of the examiner or a supplemental examiner. You may not refer to notes or any other written material during the trial. Talking is not prohibited, however the examiner is not allowed to respond to any questions, or offer coaching of any kind.

Immediately following completion of each trial, the examiner will tell you whether you passed or failed that trial. If a failure has occurred, the examiner will tell you which steps were performed or described (for verbal steps only) incorrectly. The examiner will not stop the trial at the point an error is made, or express approval or disapproval (verbal or nonverbal).

If, during a trial, you feel that you have made an error, you may request to suspend that trial and immediately begin the procedure over. If a trial is suspended voluntarily, it will not be counted as a failure of the trial. You may voluntarily suspend one trial per procedure.

The performance exam is physically demanding. If you feel you are risking personal injury in performing, or are unable to perform the procedures in the performance exam due to diminished physical capacity or physical disability, please inform the sponsoring group in advance of the examination date so alternative arrangements may be made.

<u>Appeals</u>

All ACI certification policies, procedures, requirements, and examination materials have been developed through the voluntary consensus process and carefully reviewed for accuracy and fairness. However, an appeal process is available in the event that you feel that some aspect of the examination administration is unclear, incorrect, or unfair (see **Challenges** for examination content disputes). If a complaint is found to be valid, ACI will adjust grades accordingly.

The first level of appeal rests with the examiner. If the examiner cannot satisfy your complaint, appeals are handled in order by the following people or groups:

- Sponsoring Group
- Director of Certification
- Certification Appeals Committee
- Adhesive Anchor Installer Certification— Committee C680
- Certification Programs Committee

Note: Members of the Certification Appeals Committee include the Director of Certification, the Chair of the Certification Programs Committee, and the Chair of Committee C680.

Appeals submitted directly to ACI must be received, in writing, within 60 days of the receipt of the examination at ACI Headquarters. Appeals that are not made during the examination session will not be considered.

Re-Examination

Failure of either exam by the criteria cited previously will require re-examination on the failed exam. A re-examination on either exam may be taken at any time, together or independently of one another; however, both must be passed within one year to be considered for certification.

Failure of a performance examination on one or more of the required parts will require re-examination on the entire performance exam. All portions of the performance exam must be passed at the same time. If the re-examination is not taken within one year of passing the written exam, the entire written exam must also be retaken.

It is your responsibility to request re-examination. To do so, contact the examiner and/or sponsoring group to arrange a convenient time, location, and appropriate fee. If you require further assistance, contact the ACI Certification Department.

Recertification

Your ACI certification as an Adhesive Anchor Installer expires five years from the date that you complete the certification requirements. Recertification is granted to an Adhesive Anchor Installer who successfully completes the then-current written and performance examinations.

Americans with Disabilities Act (ADA)

If you have a disability and are qualified to do a job, the ADA protects you from job discrimination on the basis of your disability. To be protected under the ADA, you must have a substantial—as opposed to a minor—impairment. (A "substantial" impairment is one that significantly limits or restricts a major life activity such as hearing, seeing, speaking, walking, breathing, performing manual tasks, caring for oneself, learning or working.) If you have a disability, you must also be qualified to perform the essential functions or duties of a job, with or without reasonable accommodation, in order to be protected from job discrimination by the ADA.

ACI Certification programs are ADA-compliant. Individuals who are seeking certification and require reasonable accommodations must provide ACI Certification with a letter from their physician that outlines the recommended accommodations. ACI will review the request and formulate a recommended course of action for the Sponsoring Group (Testing Center) to follow. All requests are handled on a case-by-case basis and must be made allowing enough time for evaluation and appropriate action by ACI.

To contact ACI Certification, call (248) 848-3790 or email aci.certification@concrete.org.

JOB TASK ANALYSIS

Preparing for Installation – Chapter 2

W	Verify that adhesive is suitable for the intended application (dry, water-filled, submerged, overhead)
W	Review manufacturer's printed installation instructions (MPII)
W	Review material safety data sheet (MSDS)
W	Select appropriate personal protective equipment (PPE)
W	Verify concrete temperature falls within range for selected product
W	Evaluate concrete condition, age, cracks, expansion joints, thickness, etc. (Chapter 4)
W	Verify and layout anchor locations according to specifications

Drilling Anchor Holes – Chapter 3

W	Adjust equipment components according to specifications
W	Take appropriate action if drilling hits reinforcing steel or other obstructions
P	Drill hole perpendicular to concrete
W & P	Determine proper depth, diameter, and rod size for hole according to specifications
W & P	Select appropriate drilling machine and bits as per manufacturer specifications
W & P	Verify appropriate dust extraction system and accessories are available and working
W & P	Verify that hole depth and diameter meet pre-determined specifications

Cleaning Anchor Holes – Chapter 4

W	Determine appropriate cleaning method per manufacturer instructions
W	Remove water-concrete particle slurries
P	Visually inspect debris coming from hole during drilling
W & P	Select appropriate equipment to clean holes
W & P	Remove debris from drill hole using appropriate cleaning method
W & P	Engage dust extraction system with accessory to capture dust during blow-out of drilled hole

Injecting Adhesive Using Cartridge Systems – Chapter 5

W	Determine appropriate adhesive working (gel) time
W & P	Verify adhesive expiration date
W & P	Identify cartridge storage temperature requirements
W & P	Verify that proper nozzle is selected with complete mixing element
W & P	Assemble adhesive cartridge and nozzle per MPII
W & P	Confirm that hole is clean prior to injecting adhesive
W & P	Insert adhesive assembly into dispenser
W & P	Select equipment suitable for installation location (horizontal to overhead)
W & P	Discard initial adhesive and confirm proper mixing
W & P	Inject adhesive per MPII (rate, application, method, avoid air entrapment)
W & P	Determine minimum depth of adhesive fill

Installing Anchors – Chapter 6

W	Identify most appropriate installation technique per MPII
W	Select the anchor appropriate for the adhesive system in use
W & P	Inspect anchor element for contaminants and clean if needed
W & P	Inspect anchor element for damage and replace if needed
W & P	Insert anchor element into borehole per MPII
W	Take corrective action if air pockets are noted during installation
W & P	Verify accurate final embedment depth based on length of protruding anchor element
W & P	Verify that adhesive fills hole uniformly around the anchor, approximately flush with surface
W	Ensure that anchor remains undisturbed until adhesive is fully cured

Installing Adhesive Capsule Systems – Chapter 7

W	Verify adhesive expiration date
W	Identify capsule storage temperature requirements
W	Confirm that material has not solidified inside capsule
W	Confirm that hole is clean prior to capsule insertion
W	Insert capsule into hole per the MPII

EDUCATIONAL MATERIALS

Reported by ACI Committee E905 - Training Programs

Thomas O. Malerk, Chair

David J. Akers	N. Mike Jackson	Christopher J. Robinson
Stephan A. Durham	Cecil L. Jones	William Rossi
Frances T. Griffith	Jon I. Mullarky	Ronald E. Vaughn

Developed by Committee E905 – Task Group on Adhesive Anchor Installer

Thomas O. Malerk, Chair

T.J. Bland	Christopher La Vine	John F. Silva
William Dubon	Lee W. Mattis	Mark Timmerman
Werner A. F. Fuchs	Donald F. Meinheit	J. Bret Turley
	James Surjan	

Chapter 1: Introduction

This document provides a general overview of the information an adhesive anchor installer needs to know to successfully install an adhesive anchor. This workbook includes theory on how adhesive anchor systems work in concrete as well as good field practices. An adhesive anchor system is made up of an adhesive, an anchor, installation equipment, the manufacturer's printed installation instructions (MPII), and material safety data sheet (MSDS). The adhesive chemistry can be an epoxy or non-epoxy, and the selection of the adhesive type depends on installation and usage conditions of the adhesive anchor. An anchor is typically a continuously threaded rod or deformed reinforcing bar that mechanically interlocks with the adhesive to transfer load into concrete through direct surface bonding at the hole interface with the concrete substrate.

Each manufacturer's adhesive anchor system has product specific instructions for installation that must be followed to ensure the anchor system reaches its full strength (or load bearing capacity). The intent of this document is to provide generic guidelines that help the installer understand and properly execute the installation procedures they will encounter in various MPII.

Chapter 1 of this document provides an introduction to the various types of concrete anchors, their components, and their behavior. Chapter 2 provides an overview of what the installer should expect to find in the MPII and MSDS, a brief overview of personal protective equipment (PPE), and typical information that may be found on package labeling. Chapters 3 through 7 discuss the tools, materials, and procedures for adhesive anchor installation. Chapter 8 covers common installation problems that may be encountered and ways to prevent or correct them.

Types of Anchors

Anchor systems are typically used to connect two parts of a structure together or to connect a nonstructural item to the structure. As shown in Fig. 1.1, the connection includes three main components, the attachment (typically a piece of steel), the anchor, and the substrate material (typically concrete or masonry into which the anchor is secured). Several terms are used in this industry to refer to an anchor, including anchor element, anchoring element, anchor rod, and threaded anchor rod. Throughout this document, the term *anchor* will be used.

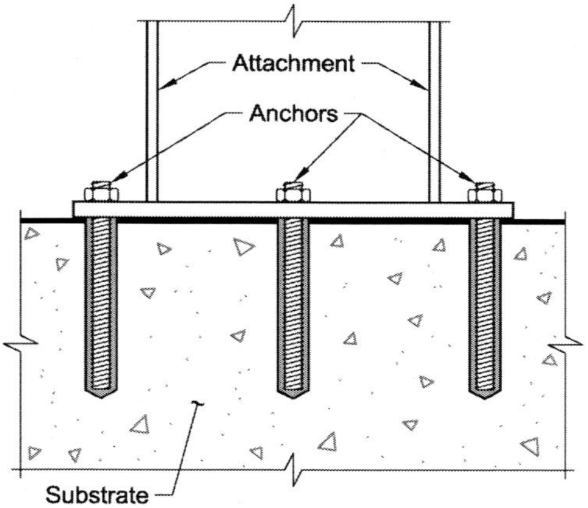

Fig. 1.1: Illustration of a connection using adhesive anchors. The connection includes an attachment, anchors, and substrate

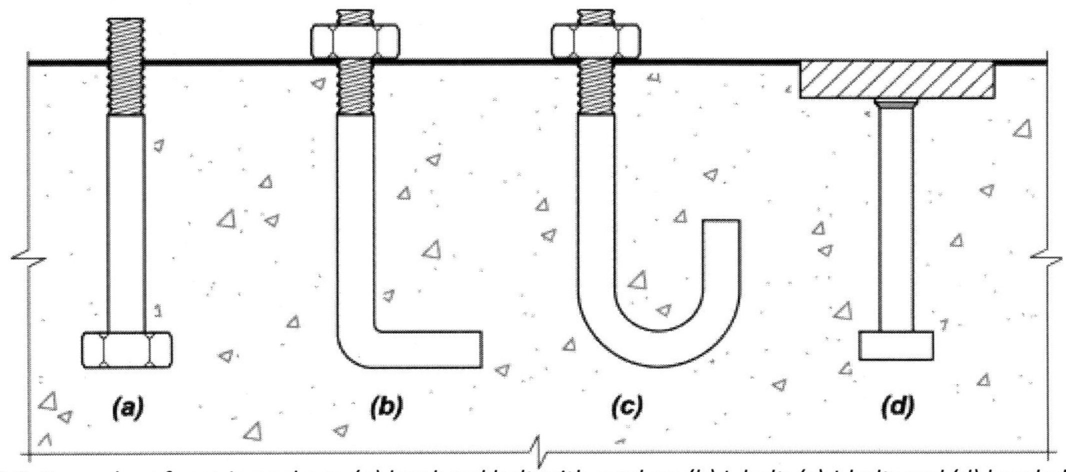

Fig. 1.2: Examples of cast-in anchors: (a) hex head bolt with washer; (b) L-bolt; (c) J-bolt; and (d) headed stud

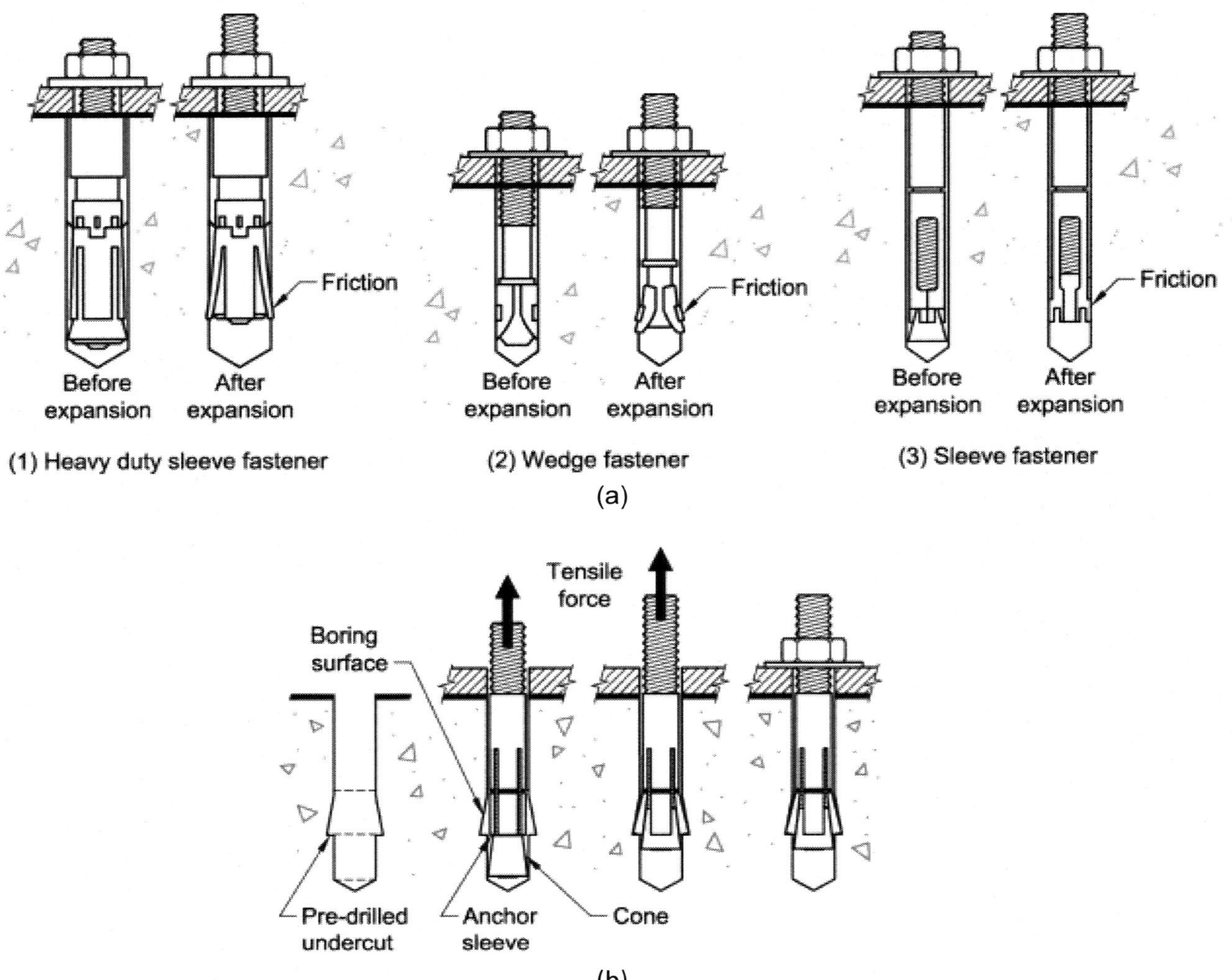

Fig. 1.3: Examples of common types of post-installed mechanical anchors: (a) mechanical expansion anchors that transfer load to the concrete through friction; and (b) an undercut anchor that transfers load to the concrete through bearing

In new construction, anchors are often headed studs, threaded rods, deformed reinforcing bars, or bolts that are cast in the concrete before it hardens. These are known as cast-in anchors. Examples of cast-in anchors are shown in Fig. 1.2. Further discussions will address post-installed anchors that are used after concrete has hardened and provide new construction anchor solutions without major demolition.

Anchors that are installed some period of time after the concrete has hardened are known as post-installed anchors. The first post-installed anchors were mechanical anchor systems. They have been used since the early 1900s to connect items to existing supporting structures and are commonly and extensively used today. Grouted anchor systems soon followed the early mechanical anchor systems. These systems use cement type products to bond to the concrete.

Post-installed mechanical anchors can be used in a variety of construction applications, especially when a connection between a new item and an existing support is required, providing economic and structural solutions.

Post-installed mechanical anchors transfer load to the concrete or masonry through either friction or bearing between the anchor and the sides of a predrilled hole in the concrete. Anchors that transfer load through friction often do so through an expansion mechanism that presses a portion of the anchor wedges against the concrete, creating friction applied using torque. They are often referred to as *expansion anchors*. Several examples are shown in Fig. 1.3(a). Post-installed anchors that transfer load through bearing are installed in holes in the concrete where a portion of the hole near the end has been expanded using a special undercutting bit, as shown in Fig. 1.3(b). When the anchor is expanded, it produces a mechanical interlock with the concrete locking the anchor in

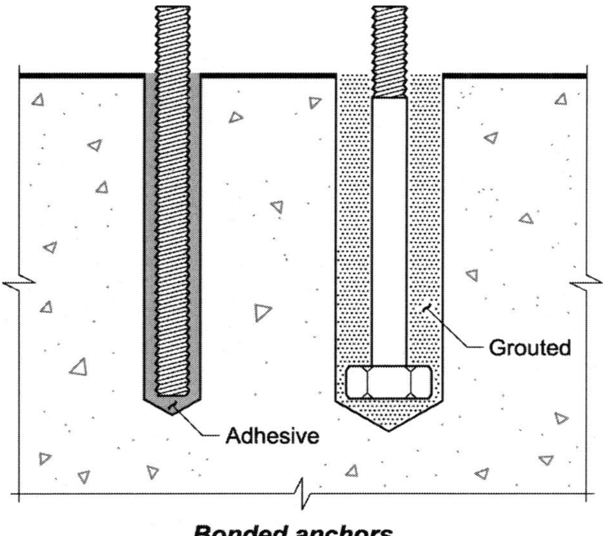

Fig. 1.4: Post-installed bonded anchors can use adhesive or grout as the bonding material

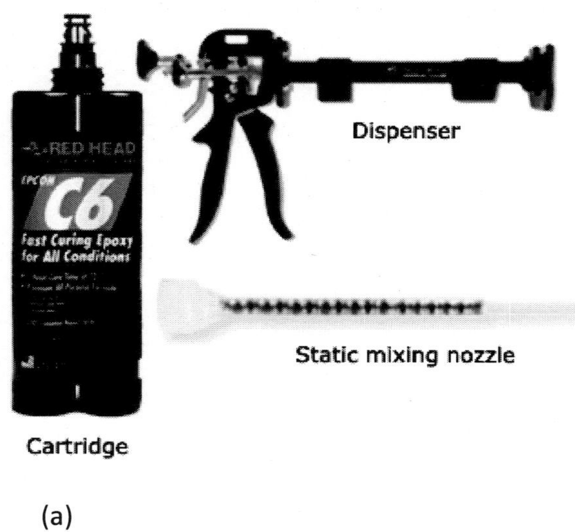

(a)

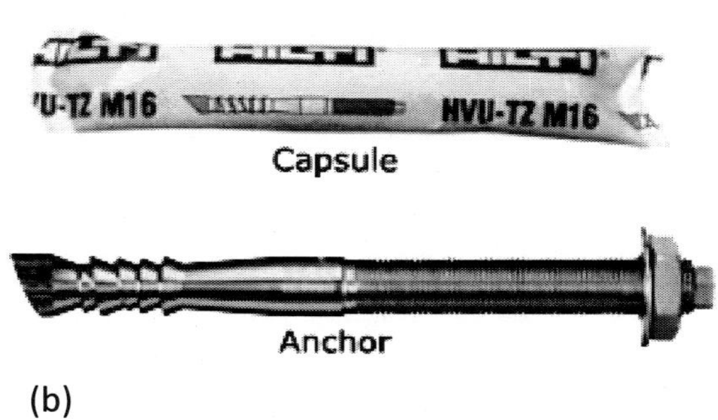

(b)

Fig. 1.5: Examples of adhesive anchor systems: (a) a cartridge type adhesive system; and (b) a capsule type adhesive system where the adhesive is contained within a capsule that is inserted into the hole and mixed as the anchor is driven into the hole. Also included in both systems is the MPII and drill (not shown) (Photographs courtesy of ITW Redhead and Hilti)

place at time of installation. This type of anchor is often referred to as an *undercut anchor*.

A *bonded anchor* is another type of post-installed anchor. It connects the anchor to the concrete using an adhesive or grout, as shown in Fig.1.4. Adhesives are quick setting polymeric materials whereas grouts are typically cement based and activated by water. With a bonded anchor, load is transferred through the adhesive or grout from the anchor to the concrete or masonry. It works by keying the adhesive or grout into the embedded depth of the anchor and hole surface to provide continuous load transfer.

An adhesive anchor system is a type of bonded anchor that uses a polymer material, epoxy or acrylic, as the bonding agent. These systems require low air entrapment and clean holes. They became popular in the 1980s when adhesive technologies improved. Although there are several types of adhesive anchors used in construction, the American Concrete Institute (ACI) Adhesive Anchor Installer certification program currently addresses only cartridge and capsule type adhesive anchor systems, where premeasured amounts of resin and hardener are mixed to form a "glue." Examples of current types of systems are shown in Fig. 1.5.

In a cartridge adhesive anchor system (Fig. 1.5(a)), the adhesive components are contained in a cartridge and dispensed using a special dispensing tool and nozzle (similar to dispensing caulk from a caulking dispenser) that are unique to each product. The nozzle contains a section called a static mixing element that completely mixes the components. The nozzle limits the amount of air trapped in the adhesive by allowing the adhesive

to be dispensed at the bottom of the hole. In a capsule system (Fig.1.5 (b)), all of the components of the adhesive are contained in a capsule that is inserted into the hole. The adhesive is released from the capsule and mixed as the anchor is driven into the hole. The anchor for both types of systems is typically a continuously threaded rod, but may also be a deformed reinforcing bar or other special anchor as shown in Fig. 1.5(b). For capsule systems, the anchor typically needs to be modified with a chisel point at one end.

Unlike post-installed mechanical anchors that function through friction or bearing, adhesive anchors connect to the concrete or masonry primarily through the bond generated by the adhesive to both the concrete and the anchor (Fig. 1.6). This bond is present over the length of the entire embedment depth, which typically ranges from 4 to 20 times the anchor diameter. Additionally,

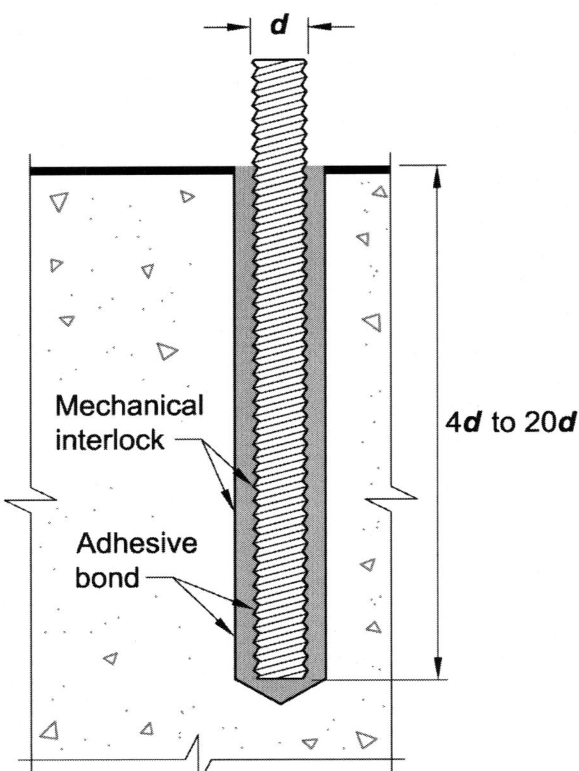

Fig. 1.6: Adhesive bonds and mechanical interlock are created when an adhesive anchor system is used in concrete. One adhesive bond occurs between the concrete and adhesive. Another occurs between the adhesive and anchor. Mechanical interlock develops between the anchor and the adhesive and can develop between the concrete and adhesive if the sides of the hole are rough

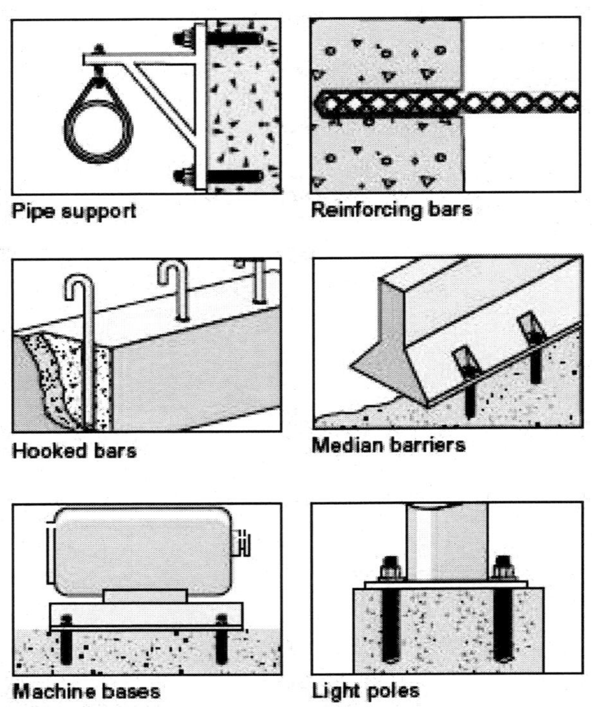

Fig. 1.7: Typical adhesive anchor applications (Courtesy of Powers Fasteners)

anchors with uneven surfaces, such as threaded rods and deformed bars, produce a mechanical interlock between the adhesive and the anchor. Adhesive anchors that are installed in holes with rough sides will also create a mechanical interlock between the concrete and the adhesive.

The benefits of using adhesive anchors, other than their high performance, are that they are post-installed anchors that can easily be used for project design changes, they support a variety of anchor solutions, and installation requires only a moderate amount of equipment. Typical applications for adhesive anchors are shown in Fig. 1.7. These applications may require the adhesive anchor to be installed in any orientation including vertical down, horizontal, overhead (also referred to as upwardly inclined or vertical-up), and any angle in between.

The decision regarding which anchor system is suitable for the intended application requires consideration of a broad range of factors including hole contamination, anchor orientation, ambient temperatures, moisture residue, and weather conditions, as well as load type and duration. The performance characteristics of adhesive anchors are found in product performance reports such as evaluation service reports (ESRs) by International Code Council - Evaluation Services (ICC-ES). ICC-ES provides technical evaluations for code compliance, providing regulators and construction professionals with clear evidence that products comply with codes and standards. The selection of the proper anchor system is the responsibility of the engineer for the project. Proper installation per the MPII of these systems, by a Certified Adhesive Anchor Installer, is critical to achieving their load capacity and performance.

Importance of Proper Installation

Proper installation techniques are vital for the success of an adhesive anchor system. Figure 1.8 shows the test setup from actual sessions of the ACI Adhesive Anchor Installer certification program for the overhead adhesive injection performance examination. The tube on the left in Fig. 1.8(a) appears to be properly filled. To ensure that there are no interior voids, however, sectioning of the tubes is performed after the performance examination. The tubes are then evaluated to determine if the installer has passed or failed this portion of the examination. The tube on the right in Fig. 1.8(a) has no adhesive in the top half and the piston plug (a special device used to fill overhead holes, further description can be found in Chapter 2) is upside down and was left in the tube. This would be a failure. The passing criterion for a properly filled tube is to have the deepest two thirds completely filled with no voids as correctly shown in Fig. 1.9(a). Voids present in the shallowest one-third of the tube are not considered.

Figures 1.9(a) to 1.9(f) illustrate how difficult it can be to properly fill a hole in the overhead orientation without trapping air voids in the hole. Many installers may not even realize they are not achieving complete filling of the overhead, horizontal, and vertical down holes unless they see the results of tests such as these. Adhesives have a relatively high resistance to flow and require specialized procedures including very slow deliberate motion and specialized equipment such as a piston plug or retaining cap for proper installation.

The quality of anchor installation in the field must meet the installation quality achieved in laboratory tests used to qualify the anchor for the design engineer to have confidence that the anchor in the field will perform properly. Rigorous testing and evaluation requirements for adhesive anchor systems are detailed in ACI 355.4-11, "Acceptance Criteria for Qualification of Post-Installed Adhesive Anchors in Concrete" and ICC-ES AC308, "Acceptance Criteria for Post-Installed Adhesive Anchors in Concrete Elements." Based on the results of these tests, the anchor design values are calculated in accordance with Appendix D of ACI 318-11, "Building Code Requirements for Structural Concrete." The engineer using the values from these tests assumes in his or her design that the installation of the anchors will be performed in accordance with the MPII. Deviations from the MPII may cause the strength to vary from the design strength values determined using ACI 318-11 Appendix D. In practice this can lead to failure of the anchor.

Anchors installed in positions that range from horizontal to overhead are considered upwardly inclined installations and are subject to special requirements in ACI 318, Appendix D. Additionally, if an adhesive anchor in an overhead installation supports a sustained tension load, the code requires that the anchor be installed by a person certified in accordance with the ACI Adhesive Anchor Installer Certification program or an equivalent program with similar requirements that is acceptable to the licensed design professional. The installation of adhesive anchors must also be inspected by an inspector specially approved by the building official. These requirements are in place to prevent anchor failures that can result in

(a)

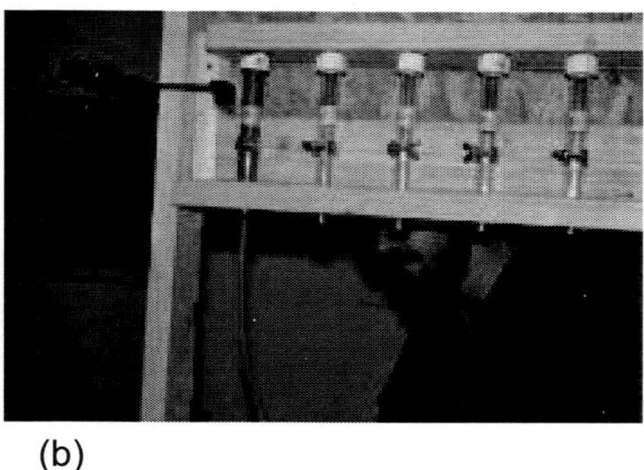

(b)

Fig. 1.8: Blind overhead tube filling setup: (a) the setup simulates an overhead installation in the field; and (b) the installer cannot see the tubes when filling them with adhesive

loss of life or damage to property.

Because there are no easily applicable non-destructive methods to confirm complete filling of the hole or to determine if air voids exist in the adhesive, special inspections of anchor installations on the job site are often required. These special inspections may consist of visual inspections during the installation process to confirm if proper drilling, cleaning, adhesive mixing, adhesive injection into the drill hole, anchor setting, and protection of adhesive anchors are performed in accordance with the MPII. Visually, an inspector can see if the hole has been filled to the top, but this

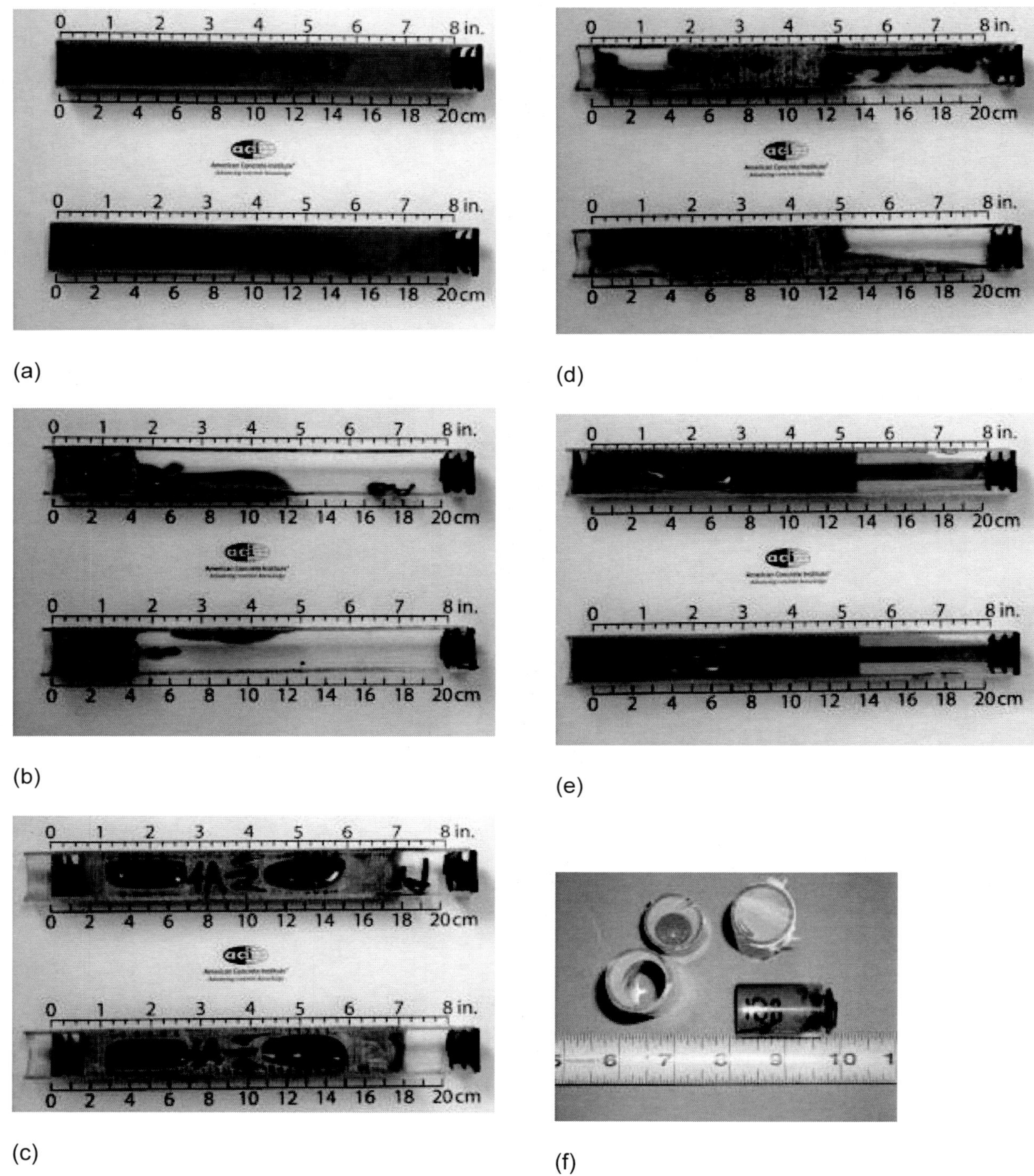

Fig. 1.9: The tubes shown in these photographs were cut for evaluation after filling by an examinee in the overhead installation for the ACI Adhesive Anchor Installer certification program. The top of the tube is positioned on the right in all of these photographs; (a) a properly-filled tube with no voids or other deficiencies; (b) the upper three-quarters of this tube is not filled, resulting in failure of the performance examination; (c) this tube has voids in the middle portion of the tube, and no adhesive at the deepest part of the hole, resulting in failure. Voids are likely due to pulling the nozzle out too quickly while dispensing the adhesive; (d) a "waterfall effect" is produced as the adhesive falls down on itself and does not fill the volume of the tube, resulting in failure. This can occur if the nozzle is initially pulled out too quickly; (e) in this tube, the piston plug is upside down and was left in the tube (blocking the installation of the anchor), resulting in failure; (f) this tube was not completely filled and had voids in the middle, resulting in failure

14

Adhesive Anchors Inspection Checklist for Concrete & Masonry

Special Inspection shall be in compliance with Section 1701 of the UBC and Section 1704 of the IBC as described below. (See Structural Drawings for Inspection requirements)

Project Name:____

Project Location:____

Weather:________________ Air Temperature:____________(°F / °C)

CODES
☐ UBC 1997
☐ IBC 2000
☐ IBC 2003
☐ IBC 2006

Seismic Zone/ Seismic Design Category

Adhesive

Product Name/Manufacturer:____

Lot No.:____

ICC-ES Report No.:____

Adhesive expiration Date: ___/___/___ ☐ Specified Dispenser ☐ Specified Mixer

Discard Initial Adhesive: __________(# trigger pulls)

Adhesive Tempature:__________(°F / °C)

Gel Time:__________ Cure Time:__________
(per manufacturers installation instructions)

Adhesive Element

Type: ☐ All-Thread ☐ Internally Threaded ☐ Torque-Controlled ☐ Rebar ☐ Other__________

Material: ☐ Standard ☐ Stainless Steel ☐ High Strength

Steel Grade/Coating:__________

Length:__________(in/mm)

Rod Diameter: ☐ 3/8" ☐ 1/2" ☐ 5/8" ☐ 3/4" ☐ 7/8" ☐ 1" ☐ 1 1/4"

Rebar: ☐ #3 ☐ #4 ☐ #5 ☐ #6 ☐ #7 ☐ #8 ☐ #9 ☐ #10

☐ #11 ☐ Other__________

Base Material

Base Material Type: ☐ NW Concrete ☐ LW Concrete ☐ Brick ☐ CMU ☐ Other__________

Base Material Strength: ☐ 2000psi ☐ 3000psi ☐ 4000psi ☐ Other__________

Base Material Thickness:__________(in/mm)

Base Material Temperature:__________(°F / °C)

Drilling & Hole Cleaning

Drill Bit Diameter:__________(in/mm) Hole Depth:__________(in/mm)
(core bit diameter, if approved)

Drill Bit Type: ☐ Carbide-Tip Drill Bit ☐ Diamond Core Bit ☐ Other__________
(ANSI B212.15-1994)   (if appropriate and allowed)

Hole Condition: ☐ Dry ☐ Water Saturated ☐ Water Filled ☐ Under Water

Hole Cleaning: ☐ Compressed Air ☐ Hand Pump ☐ Wire Brush ☐ Nylon Brush ☐ Other__________

Hole cleaning in accordance with manufacturers' printed installation instructions: ☐ Yes ☐ No

---

Application

Anchor Application: (please check all that apply) ☐ Tension ☐ Shear ☐ Overhead ☐ Other__________

Anchor insertion: ☐ Twisting motion ☐ Annular gap filled with adhesive ☐ Air Void Free Injection

Anchor Spacing:__________(in/mm) Edge Distance:__________(in/mm)

Embedment (h_{ef}):__________(in/mm) Installation Torque (if required):__________(ft-lb/Nm)

Completed by: ____________________(Signature) Date:____/____/____

____________________(Print) Company:____

____________________(Title)

Version 09.2007

Fig. 1.10: Example of a typical adhesive anchor inspection checklist for concrete and masonry applications (Courtesy of Hilti)

Fig. 1.11: Big Dig adhesive anchor bond failure. These are adhesive anchors that have pulled out. The anchors are supporting a concrete ceiling in the tunnel (Photograph courtesy of National Transportation Safety Board). The anchor pictured here was dislodged from the I-90 tunnel ceiling (Photograph courtesy of MassDOT)

does not guarantee that the entire hole was filled uniformly throughout the depth. Proof loading, a load test after the adhesive has cured, may be used to verify adequate load-carrying capacity of the adhesive anchor. This can be used as a supplement to special inspection, especially as an incentive to promote good installation techniques.

Typical special inspections require documentation of the following items. These items may vary depending on the product:

- Anchor type;
- Anchor dimensions, material, grade (for reinforcing bar and threaded rod), cleanliness;
- Drill type;
- Required drill bit diameter, including verification of carbide-tipped drill bit;
- Concrete type;
- Concrete compressive strength;
- Adhesive identification and expiration date;
- Hole dimensions;
- Hole cleaning procedures;
- Anchor spacing;
- Edge distances;
- Concrete thickness;
- Substrate and adhesive temperature at time of anchor installation;
- Actual gel time when installed anchors are not disturbed and properly inserted to bottom of hole;
- Anchor embedment;
- Adhesive mixing procedure;
- Tightening torque; and
- Adherence to manufacturer's printed installation instructions.

Figure 1.10 is an example of a form that might be used in an adhesive anchor inspection.

After initial inspection, subsequent installations performed by the same construction site personnel may not require inspection. A change in staff or products, however, requires another inspection. A large project may require continuous inspections at reasonable time intervals.

Examples of Anchor Failures

When anchors are not selected or installed properly there can be very serious consequences. A highly-publicized example of anchor failure occurred in 2006 when concrete panels fell from the ceiling of the I-90 Connector Tunnel Project (also known as the Big Dig Project) in Boston, MA. The National Transportation Safety Board (NTSB) investigated and found that the panels were attached to the tunnel ceiling using an adhesive anchor system that was not appropriate for the application.

The NTSB recommended that ACI "use your building codes, forums, educational materials, and publications to inform design and construction agencies of the potential for gradual deformation (creep) in anchor adhesives and to make them aware of the possible risks associated with using adhesive anchors in concrete under sustained tensile-load applications."[1] Additional findings in the NTSB report noted that the following installation errors occurred: holes were not blown out with compressed air or brushed; the holes intersected reinforcing bars, and there was poorly mixed adhesive or voids in the adhesive. Many of these mistakes may have been avoided if proper installation techniques had been used. Examples of failed connections and anchors found at the Big Dig site are shown in Fig. 1.11.

A sustained load is a load that is continuously applied to the anchor system. Polymer adhesives can creep or move under sustained load. In response to the recommendations of the NTSB, Appendix D of ACI 318-11 requires that adhesive anchors shall be qualified for use in these types of applications. Additionally, Appendix D requires that installation of adhesive anchors horizontally or upwardly inclined (overhead) that are intended to support sustained tension loads shall be performed by personnel certified by an applicable certification program such as the ACI Adhesive Anchor Installer Certification program.

Adhesive Anchor Failure Modes

Loads that anchors resist can be classified into two main categories: tension and shear. As shown in Fig. 1.12, tension is a load that acts perpendicular to the surface of the concrete or parallel to the axis of the anchor. Shear is a load that acts parallel to the surface of the concrete or perpendicular to the

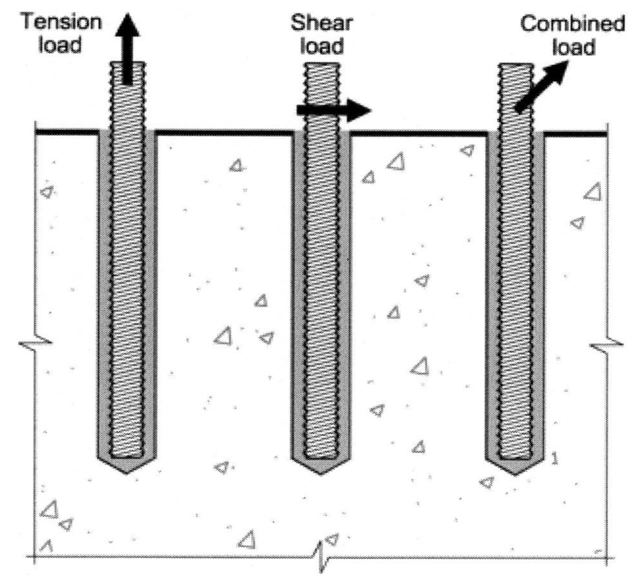

Fig. 1.12: Tension, shear, and combined loads can be applied to adhesive anchors

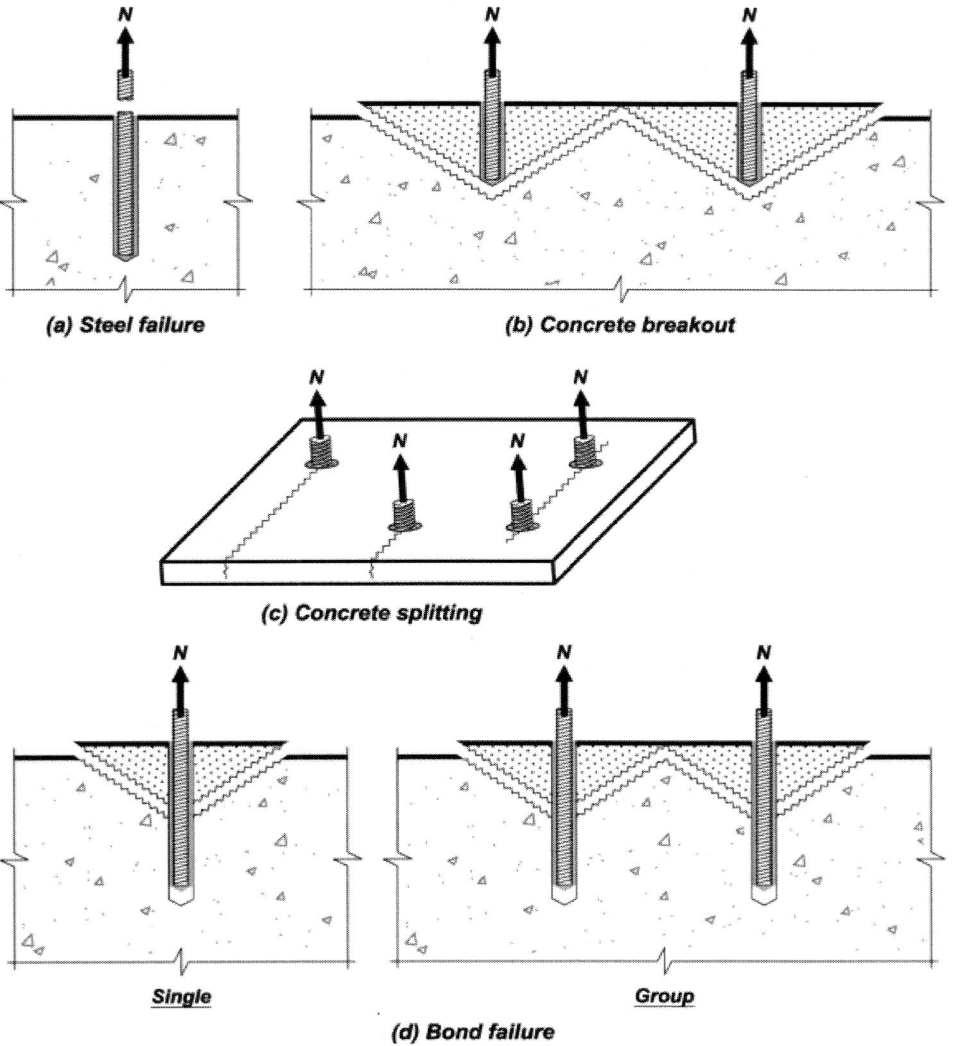

Fig. 1.13: Adhesive anchor failure modes for anchors with tensile loading

Fig. 1.14: Example of a typical steel failure (Photograph courtesy of Wiss, Janney, Elstner Associates, Inc.)

axis of the anchor. Often, anchors are subjected to a combination of tension and shear loads, as shown in Fig. 1.12.

Tightening a nut on an adhesive anchor will create tension in the anchor. This should only be done after the recommended curing time for the adhesive has passed.

Adhesive anchor failure modes in tension are illustrated in Fig. 1.13. The failure modes include steel failure, concrete breakout failure, concrete splitting failure, and adhesive bond failure. Anchor failure occurs when the applied load exceeds the weakest of these failure modes.

Steel failure occurs when the applied load is greater than the load the steel can support. The steel failure load depends on the properties of the steel and the diameter of the anchor. Anchors with deep embedment depths installed far from edges of the concrete are more likely to fail in this manner. Of the possible ways an anchor can fail, steel failure is often preferred because stretching of the anchor before it fails can provide a warning that failure may soon occur. The failure load can also be predicted more accurately than for most other failure modes. As shown in Fig. 1.14, typical steel failures occur in the anchor just above the concrete surface.

Concrete breakout failure results in breakout of a cone-shaped piece of concrete, as shown in Fig. 1.15. In this case the adhesive anchor system is stronger than the concrete. The anchor and adhesive bond will remain intact. Anchors with shallow embedment depths typically fail in this mode.

Concrete splitting failure, shown in Fig. 1.16, occurs if the anchor is positioned too close to an edge or to another anchor. Drilling holes for

an adhesive anchor can cause very small cracks to form around the hole. Therefore, the ACI 318 design code has a minimum edge distance requirement for adhesive anchors of $6d_a$, where d_a is the nominal diameter of the anchor, unless special requirements are met.

Bond failures (Fig. 1.17) occur for a variety of reasons. Frequently, a small breakout of concrete may occur at the concrete surface in this failure mode, (Fig. 1.17(b)). If the applied load exceeds the strength of the bond between the adhesive and the concrete or the adhesive and the anchor, the system may fail by the anchor pulling out of the hole. Bond failure may also be referred to as a pull-out failure.

For adhesive anchor systems, bond failures may occur when there is minimal to no adhesive securing the anchor to the drill hole or when the adhesive cannot fully transfer the load to the concrete. This may also occur if the hole is not cleaned properly, the anchor is oily or dirty when installed, the adhesive is not properly mixed, the adhesive is injected with excessive air pockets, or the adhesive does not cure properly or otherwise fails to set. The strength of the bond can be significantly affected by the actions of the installer.

Adhesive anchor failure modes in shear include steel failure, concrete pryout failure, and concrete breakout failure (Fig. 1.18).

The strength at which steel failure of an anchor in shear occurs depends on the properties of the anchor material and the physical dimensions of the anchor. An example of an anchor that has undergone a shear failure is shown in Fig. 1.19. This type of failure is often seen with deep embedment depths located far from the edges of the concrete.

Fig. 1.15: Concrete breakout failure (Photograph courtesy of IWB, University of Stuttgart)

Fig. 1.16: Concrete splitting failure (Photograph courtesy of Powers Fasteners)

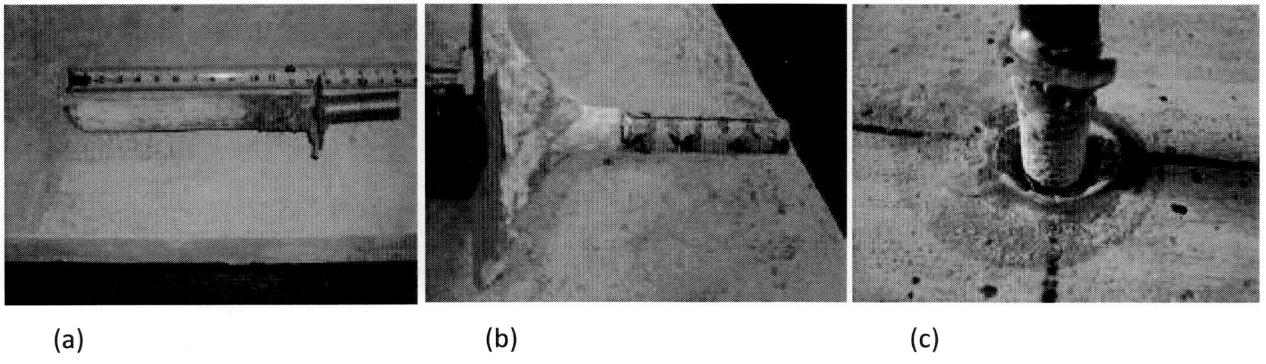

(a) (b) (c)

Fig. 1.17: Adhesive anchor bond failures: (a) An anchor that experienced bond failure between the adhesive and the concrete (Photograph courtesy of Wiss, Janney, Elstner Associates, Inc.); (b) An anchor that experienced bond failure between the adhesive and the reinforcing bar (Photograph courtesy of Wiss, Janney, Elstner Associates, Inc.); (c) An anchor that experienced bond failure between the adhesive and the reinforcing bar (Photograph courtesy of ITW Red Head)

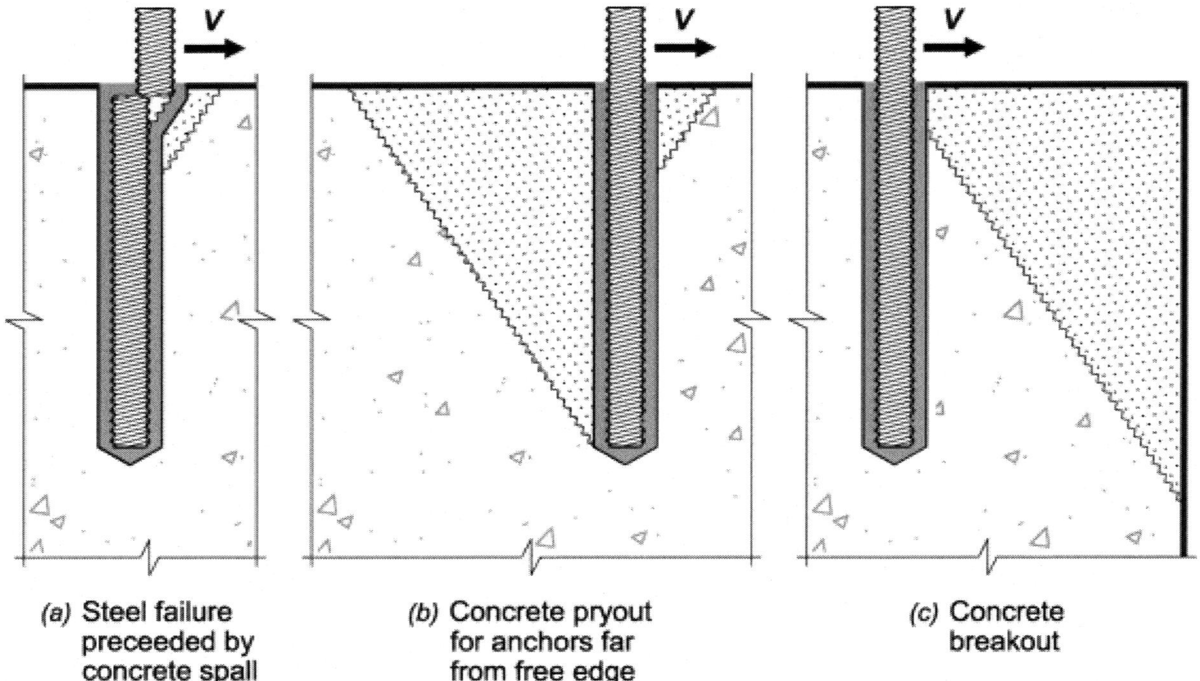

Fig. 1.18: Adhesive anchor failure modes in shear

Fig. 1.19: Shear failure of an adhesive anchor (Courtesy of Wiss, Janney, Elstner Associates, Inc.)

Fig. 1.20: Concrete breakout failure of an anchor located near a free edge (Courtesy of IWB, University of Stuttgart)

Concrete pryout occurs at very shallow embedment depths and corresponds to the formation of a concrete spall behind a short, stiff anchor. The spall occurs on the side of the anchor that is opposite the direction of the applied shear force. This type of failure is illustrated in Fig. 1.18(b).

Concrete breakout failure in shear occurs when the anchor and a volume of concrete between the anchor and an edge separate from the remaining concrete in the member, as shown in Fig. 1.20. This failure typically occurs because there is little concrete in front of the anchor.

Designers of anchorage connections are required to evaluate the strength for all potential failure modes to establish the design for the anchorage system. The selection of an anchor system will be governed by the code design criteria, service conditions, eccentric loading, and installation requirements.

Chemistries and Working Principles of Adhesives

A key component of the adhesive anchor system is the adhesive. The adhesive is necessary to transfer the load from the anchor that is embedded into the concrete.

If the adhesive is not able to achieve its designed performance behavior, an anchor failure is likely. Causes of adhesive failure can include internal and external factors. The internal factors are chemical formulation and the manufacturing and packaging processes. The installer has no control over these factors.

The external factors include installation factors and in-service conditions as described briefly below. The anchor installer and designer have some level of control over these factors. Installation factors include mixing time, age of the product, and storage temperatures. In-service conditions include curing, initial loading, design loads, environmental conditions, resistance to solvents and fire exposure.

Both epoxy and non-epoxy based adhesives are used in adhesive anchoring systems. Adhesives have different bond strengths. All bond strengths are affected by the condition of the anchor, the hole surface, and the depth of the anchor into the concrete.

An epoxy is a copolymer, which is a material formed from combining two different components. Resin and hardener are the two components used to create an epoxy. Aggregates (usually fine sand) or other fillers are added to the epoxy during production to improve strength or resistance to elevated temperature. Epoxy adhesives can be fast or slow setting, flexible or rigid, and have heat and chemical resistance. The chemical reaction can generate heat, so recently mixed epoxy should be handled with caution.

The process of polymerization (curing) of the adhesive is affected by temperature. The temperature of the adhesive and concrete serving as the substrate are the primary factors that control this process. When the temperature of the adhesive components is warm, the molecules in the components are freer to move quickly within the mixed material and link up faster. This behavior will reduce the gel time, or working time, which is the amount of time between first mixing the adhesive and when it becomes a non-flowing material.

For certain products the gel time under warm temperature concrete conditions might be as short as one minute. In this case the installation of anchors with large diameter or embedment depth may be difficult, since the gel time could be exceeded and adhesive be hardened before the anchor is inserted into the hole.

When the gel time is exceeded, the anchor is not permitted to be inserted into the adhesive. When the temperature of the components is cold, the molecules in the components move more slowly within the mixed materials and link up slower. This behavior will extend the gel time. If the components are too cold, the reaction may become dormant (stop). Dormancy may cause the anchor to fail because the adhesive has not adequately cured to achieve proper strength or bond to the concrete and anchor.

Adhesive components are mixed in different volume ratios. Each adhesive product should be mixed at the manufacturer's specified ratio. Adhesives are available in cartridge, capsule, or bulk systems. For each system, the components are separately premeasured and then combined. The cartridge system components are mixed through a special static mixing nozzle as they are dispensed (Fig. 1.21).

A capsule system contains both resin and hardener components in an appropriate ratio that mix when punctured by the insertion and in most cases the rotation of the anchor. Either foil or glass capsules (Fig. 1.22) are placed in the hole and then the component mixing action is provided by the anchor itself. If the hole depth requires, and is permitted by the MPII, multiple capsules can be stacked in the hole. It is not permitted to use only a portion of a capsule because the mixing ratios will be out of balance.

Bulk systems are not addressed in the ACI Adhesive Anchor Installer certification program.

Non-epoxy adhesives (including acrylics, vinyl esters, esters, and hybrid systems of organic and inorganic bonding agents) can harden through different chemical reactions. The resins and hardeners are modified in certain ways to achieve special characteristics for the intended use of the adhesive anchor.[2] For example, in an ester resin or acrylic, the mix ratio is not as critical and colder temperatures do not have as significant of an impact on the reaction of the resins. Maintaining proper temperature conditions for adhesive mixing, however, is still required.[3]

Factors Influencing Bond Strength

Bond strength is influenced by a number of factors including:

- Drilling method;
- Hole cleaning;
- Mixing;
- Installation in dry, damp, wet, or submerged concrete;
- Slurry, ice, or snow in the hole;
- Condition/age of the adhesive;
- Hole diameter;
- Embedment depth;
- Temperature;
- Curing; and
- Securing the anchor.
-

The installer has direct control over many of these factors. They will be discussed in more detail as we work through the steps involved in

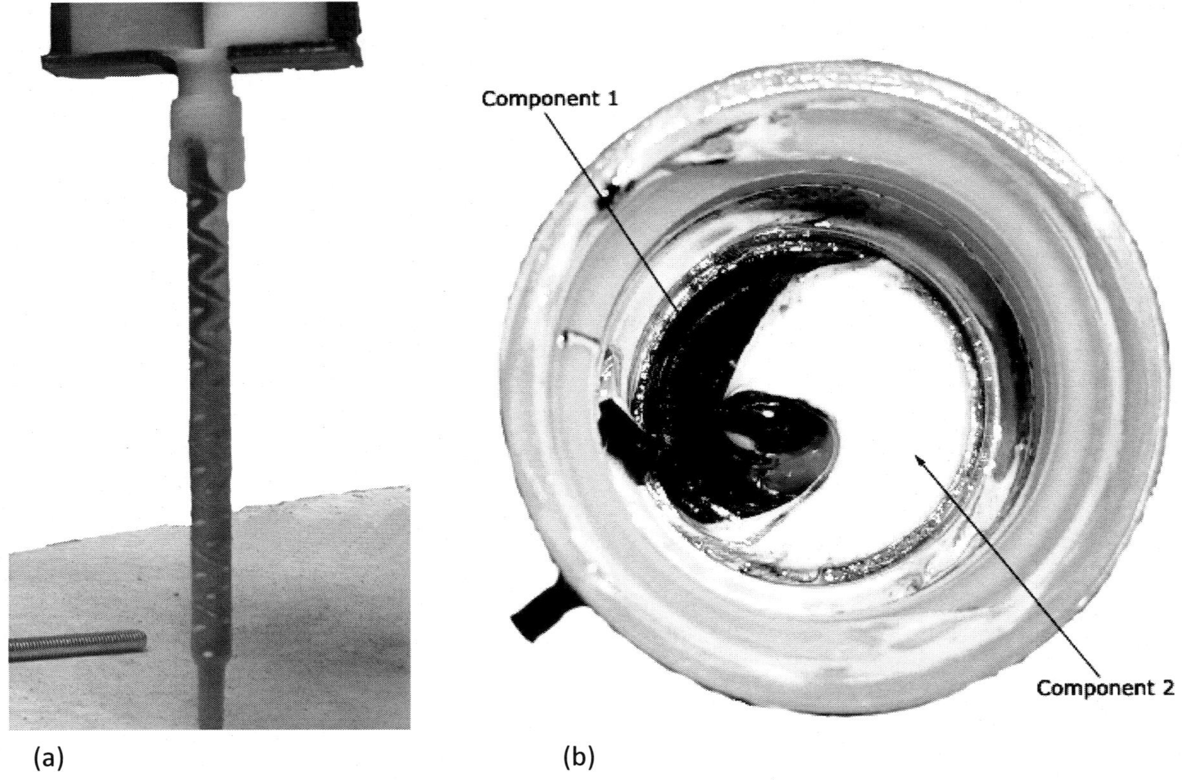

Fig 1.21: Mixing of the adhesive in an (a) adhesive cartridge system occurs as the two components travel through the mixing element. Mixing nozzles are unique to each product. (b) This is a view of the inside of the mixing nozzle with the two components just starting to be mixed

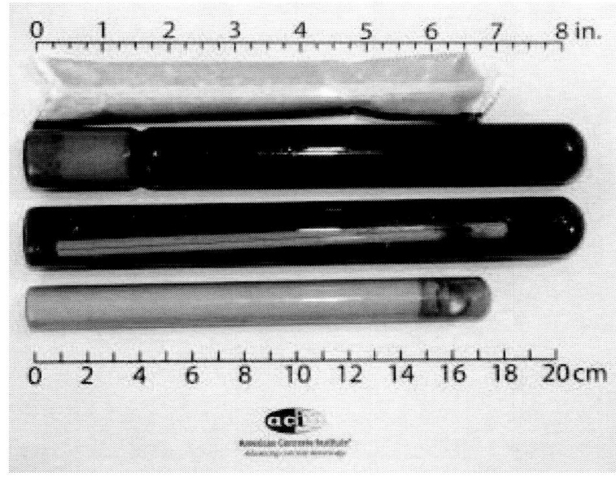

Fig. 1.22: Examples of foil and glass capsule adhesives

the installation process. By following the MPII and employing good general practices, installed anchor systems will perform as the manufacturer intends and have an opportunity to achieve the designed bond strengths. The engineer bases the anchorage design on these bond strengths.

Potential exposure of the anchor connection to freezing and thawing, chemicals, sustained load conditions, seismic and fatigue conditions, and cracked concrete are factors considered by the engineer during the design and selection process of the anchor systems. The installer has no direct control over these factors.

Cited References

[1] National Transportation Safety Board. "Ceiling Collapse in the Interstate 90 Connector Tunnel, Boston, Massachusetts, July 10, 2006," Highway Accident Report NTSB/HAR-07/02, National Transportation Safety Board, Washington, DC, July 2007, 120 pp.

[2] Eligehausen, R., and Fuchs, W., "Adhesive Anchors – Requirements for Their Reliable Use in Concrete Construction," ACI Special Publication, Vol. 283, No. 1, 2012, pp. 1-8.

[3] Powers Fasteners, "Adhesive Anchoring Systems Design Manual," January 2011 Edition.

Chapter 1

1. An adhesive anchor system is made up of an _________________________________.

   A. anchor, adhesive, cleaning equipment, and installation equipment
   B. attachment, anchor, and substrate
   C. adhesive, anchor, drilling equipment, cleaning equipment, installation equipment, MPII, and MSDS
   D. adhesive, drilling equipment, cleaning equipment, installation equipment, and MPII

2. An anchor connection includes an attachment, an anchor, and the substrate.

   A. True
   B. False

3. Adhesive anchors are cast in concrete before the concrete hardens.

   A. True
   B. False

4. What piece of equipment has a static mixing element?

   A. Drill
   B. Air compressor
   C. Mixing nozzle
   D. Dispenser

5. Adhesive anchor failure in tension does NOT occur by _______________________.

   A. steel failure
   B. concrete breakout failure
   C. concrete splitting failure
   D. adhesive bond failure
   E. concrete pryout failure

6. Adhesive anchor failure modes in shear include: _______________________________.

   A. steel failure, concrete pryout failure, and concrete breakout failure
   B. adhesive bond failure, steel failure, and concrete breakout failure
   C. concrete splitting failure, steel failure, and concrete breakout failure
   D. adhesive bond failure and concrete splitting failure

7. Installation factors that influence the strength of the adhesive anchor bond include: _____________.

   A. drilling of the hole following the MPII
   B. cleaning the hole following the MPII
   C. injecting the adhesive following the MPII
   D. all of the above
   E. none of the above

8. Adhesives should be stored _________________.

   A. at room temperature
   B. in a freezer
   C. as directed by the MPII
   D. as directed by the MSDS

9. Gel time is _______________________________.

   A. the amount of time between first mixing the adhesive and when it becomes a non-flowing material
   B. increased when the temperatures of the concrete are colder
   C. decreased when the temperatures of the concrete are warmer
   D. all of the above
   E. none of the above

10. Proper installation of adhesive anchor systems is important to ensure that the system achieves its maximum performance and to prevent anchor failures.

   A. True
   B. False

11. ACI ____, Appendix D requires that installation of adhesive anchors horizontally or upwardly inclined that support sustained tension loads are performed by certified personnel.

   A. 318
   B. 355.2
   C. 355.3
   D. 355.4

12. A post-installed anchor is ______________________________.

   A. installed in an existing concrete member
   B. installed before the fresh concrete is placed
   C. a hex head bolt
   D. a J- or L-bolt

13. A tension load is ________________ to the axis of the installed anchor.

    A. perpendicular (90°)
    B. angled
    C. parallel (the same)
    D. leaning

14. An adhesive anchor may fail if the applied load exceeds the strength of the bond between the adhesive and the concrete or the bond between the adhesive and the anchor.

    A. True
    B. False

15. An anchor can also be called ____________________.

    A. an anchor element
    B. an anchor rod
    C. a threaded rod
    D. an anchoring element
    E. all of the above
    F. none of the above

16. Rigorous testing and evaluation requirements for adhesive anchor systems are detailed in ACI ___, "Acceptance Criteria for Qualification of Post-Installed Adhesive Anchors in Concrete."

    A. 318
    B. 355.2
    C. 355.3
    D. 355.4

SECTION 2

Chapter 2: Information for Proper Installation

Proper installation of an adhesive anchor is very critical relative to the anchor achieving its load carrying capabilities. To properly install an adhesive anchor, it is important to follow the MPII. Additional factors influencing the installation include the condition of the adhesive, anchor, installation equipment, and the concrete.

Manufacturer's Printed Installation Instructions (MPII)

Manufacturers must provide installation instructions for each adhesive anchor system they produce commercially. These instructions are commonly referred to as the MPII. Prior to installation, the installer should thoroughly read and understand these instructions. Figures 2.1, 2.2, 2.3, and 2.4 show MPII from actual manufacturers (these are included for instructional use only, please refer to the MPII for the actual product to be installed). At a minimum, the MPII will include directions for:

- Drilling the holes;
- Cleaning the holes of dust, drill fines, slurry, and debris;
- Preparing the installation equipment;
- Filling the hole with adhesive;
- Installing the anchor;
- Gel time of the adhesive;
- Curing the adhesive;
- Anchor sizes; and
- Handling and storage procedures for the adhesive.

These bullet items are unique to each adhesive anchor system and may be text, illustrations, or a combination of the two. Questions regarding the suitability of a product for a particular installation can usually be answered by checking the MPII or package labeling. If the MPII or package labeling indicate that a product is not suitable for the specific installation, the anchor installer needs to notify their supervisor of the situation and wait for further instructions before proceeding with the installation. For example, if an adhesive anchor is to be installed in a submerged application and the packaging of the adhesive system provided indicates that the system is not suitable for submerged application, the anchor installer needs to immediately stop work and notify their supervisor before any further installation work is completed.

Drilling holes: The portion of the MPII addressing drill-hole preparation includes the type of drill, type and size of drill bit, and cleaning method necessary to create a drill hole that is the proper depth, orientation, and cleanliness required for a successful anchor installation. Drill-bit sizes (these may be called setting parameters, specifications and details, or setting details) for a variety of anchor sizes typically are provided in the MPII. The required drill-bit size, found in the MPII, for the anchor diameter to be installed ensures the necessary space for the adhesive to form its bond to the concrete and the anchor.

Cleaning holes: Types of brushes (steel or nylon fiber) will be specified by the manufacturer in the MPII. Brush sizes for cleaning are selected based on the size of the drill bit used. Brush sizes will be slightly larger than the diameter of the drill bit to ensure complete contact with the sides of the hole during the cleaning process. Information on the type or types of air supply (cans of compressed air, compressor, or hand-pump) required to remove drilling dust from the hole will also be specified. Some systems may also require air manifolds, or use of a vacuum. All air nozzle use should start at the deepest part of hole, moving toward the exit of the hole and not prevent the exit of debris.

Dispensing tools, adhesive injection, and curing: The adhesive, mixing nozzle, and hand powered dispensing tool are different for each manufacturers adhesive anchor system. Each system is uniquely designed to fill the hole with air free adhesive. Because of these differences, each MPII contains illustrations, text, or both describing how to assemble the dispensing equipment. Descriptions of the adhesive, safe handling, storage instructions, and manufacturer contact information may also be provided. Additionally, information on the gel time of the adhesive, which depends on the ambient temperature, concrete temperature, and cartridge temperature, needs to be considered by the installer and be available in the MPII.

Instructions for the injection and curing of the adhesive, as well as the insertion of the anchor can be found on the MPII. These directions may vary depending on the orientation of the anchor (vertical down, horizontal, and overhead). If the product is not designed for use in any horizontal

POWERS PE1000+ Instruction Card

DESCRIPTION:
PE1000+ is an easy dispensing, high strength, 100% solids epoxy anchoring adhesive which is formulated for use in anchoring applications by trained professionals. Please refer to Powers Fasteners installation instructions and MSDS for additional detailed information.

PRECAUTION:
Safety glasses and dust masks should be used when drilling holes into concrete, stone and masonry. Wear gloves and safety glasses when handling and dispensing adhesive. Do not sand the adhesive since this can create silica dust which could be inhaled. Avoid skin and eye contact, use a NIOSH-approved chemical mask to avoid respiratory discomfort. If working indoors or in a confined area, or if sensitive to adhesive odors, wash hands or other affected body parts with soap and water if skin contact occurs. Flush eyes with plenty of water and seek immediate medical attention if eye contact occurs. Move to fresh air if adhesive odor begins to cause discomfort.

IMPORTANT!
Before using, read and review Material Safety Data Sheet (MSDS).

This product contains crystalline silica and as supplied does not pose a dust hazard. IARC classifies crystalline silica (quartz sand) as a Group I carcinogen based upon evidence among workers in industries where there has been long-term and chronic exposure (via inhalation) to silica dust, e.g. mining, quarry, stone crushing, refractory brick and pottery workers. This product does not pose a dust hazard; therefore, this classification is not relevant. However, if reacted (fully cured) product is further processed (e.g. sanded, drilled) be sure to wear proper respiratory and eye protection to avoid health risk.

HANDLING AND STORAGE
Store in a cool, dry, well ventilated area at temperatures between 32°F (0°C) and 95°F (35°C). Keep away from excessive heat and flame. Keep partially used containers closed when not in use. Protect from damage. Store away from heat and light.

Note expiration date on product label before use. Do not use expired product. Cartridge temperature must be between 41°F - 104°F (5°C - 40°C) when in use. Partially used cartridge may be stored with hardened adhesive in the attached mixing nozzle. If the cartridge is reused, attach a new mixing nozzle and discard the initial quantity of the anchor adhesive as described in the setting instructions (steps #3 and #5).

Powers Fasteners, Inc.
2 Powers Lane
Brewster, NY 10509 U.S.A.
www.powers.com
P: +1 (914) 235-6300 or (800) 524-3244

3. Gel (working) times and curing times

Temperature of base material		Gel (working) time	Full curing time
41°F	5°C	180 minutes	50 hours
50°F	10°C	120 minutes	30 hours
68°F	20°C	30 minutes	10 hours
86°F	30°C	20 minutes	6 hours
104°F	40°C	12 minutes	4 hours

It is recommended that the cartridge temperature when in use does not differ significantly from the temperature of the base material.

4. Setting parameters

Table 4.1 Specifications for installation of threaded rods

Anchor property/Setting information	Nominal threaded rod size						
	3/8"	1/2"	5/8"	3/4"	7/8"	1"	1-1/4"
d = Nominal anchor rod diameter (in.)	0.375	0.500	0.625	0.750	0.875	1.000	1.250
A_{se} = Nominal area of threaded rod (in.²)	0.078	0.142	0.226	0.335	0.462	0.606	0.969
d_0 (d_{bit}) = Nominal ANSI drill bit size (in.)	7/16	9/16	11/16	7/8	1	1-1/8	1-3/8
T_{max} = Maximum torque (ft.-lb.) for A193 B7 carbon steel rod or F593 SS rod	16	33	60	105	125	165	280
T_{max} = Maximum torque (ft.-lb.) for A36/A307 carbon steel rod only	10	25	50	90	125	165	280
$h_{ef,min}$ = Minimum embedment (inches)	2-3/8	2-3/4	3-1/8	3-1/2	[illegible]	4	5
$h_{ef,max}$ = Maximum embedment (inches)	4-1/2	6	7-1/2	9	[illegible]	12	15
s_{min} = Minimum spacing (inches)	1-7/8	2-1/2	3-1/8	3-3/4	[illegible]	5	6-1/4
c_{min} = Minimum edge distance (inches)	1-7/8	2-1/2	3-1/8	3-3/4	[illegible]	5	6-1/4
h_{min} = Minimum member thickness (inches)	h_{ef} + 1-1/4				h_{ef} + d_0		

Table 4.2 Specifications for installation of deformed steel reinforcing bars

Anchor property/Setting information	Reinforcing bar size							
	#3	#4	#5	#6	#7	#8	#9	#10
d = Nominal bar diameter (in.)	3/8	1/2	5/8	3/4	7/8	1	1-1/8	1-1/4
d_0 (d_{bit}) = Nominal ANSI drill bit size (in.)	7/16	9/16	11/16	7/8	1	1-1/8	1-3/8	1-1/2
$h_{ef,min}$ = Minimum embedment (inches)	2-3/8	2-3/4	[illegible]	3-1/2	3-1/2	4	4-1/2	5
$h_{ef,max}$ = Maximum embedment (inches)	4-1/2	6	[illegible]	9	10-1/2	12	13-1/2	15
s_{min} = Minimum spacing (inches)	1-7/8	2-1/2	[illegible]	3-3/4	4-3/8	5	5-5/8	6-1/4
c_{min} = Minimum edge distance (inches)	1-7/8	2-1/2	[illegible]	3-3/4	4-3/8	5	5-5/8	6-1/4
h_{min} = Minimum member thickness (inches)	h_{ef} + 1-1/4				h_{ef} + $2d_0$			

5. PE1000+ adhesive anchor system selection table

Injection tool	Plastic cartridge system	Extra mixing nozzle
PE1000+ 13 fl. oz. manual dispenser Cat. #06295	13 fl. oz. dual cartridge w/mixing nozzle & extension tube - Cat. #05005D	PE1000+ mixing nozzle and extension tube Cat. #06293 or 06294
PE1000+ 13 & 20 fl. oz. manual dispenser Cat. #06296	20 fl. oz. dual cartridge w/mixing nozzle & extension tube - Cat. #05025D	PE1000+ mixing nozzle and extension tube Cat. #06293 or 06294

6. Adhesive Piston Plugs

Threaded rod diameter (inch)	Rebar size (no.)	ANSI drill bit diameter (inch)	Plug Size (inch)	Plastic Plug (Cat. #)	Horizontal and overhead installations
3/4	#6	7/8	7/8	08300	
7/8	#7	1	1	08301	
1	#8	1-1/8	1-1/8	08303	
1-1/4	#9	1-3/8	1-3/8	08305	
-	#10	1-1/2	1-1/2	08309	

A plastic extension tube (3/8" dia. - Cat. #06281) must be used with piston plugs.

1. Setting instructions for solid base material - For any application not covered by this document please contact Powers Fasteners (ESR-2583)

2. Hole cleaning tools - wire brushes and air blowers

Threaded rod diameter (inch)	Rebar size (no.)	ANSI drill bit diameter (inch)	Min. brush diameter, D_{min} (inches)	Brush length, L (inches)	Steel wire brush (Cat. #)	Air blowers
3/8	#3	7/16	0.475	8-3/4	08284	Hand pump (volume 25 fl. oz.) or compressed air nozzle (min. 90 psi)
1/2	#4	9/16	0.600	8-3/4	08285	
5/8	#5	11/16	0.735	7-7/8	08286	
3/4	#6	7/8	0.920	7-7/8	08287	Hand pump - Cat. #08280
7/8	#7	1	1.045	11-7/8	08288	Compressed air nozzle only (min. 90 psi)
1	#8	1-1/8	1.175	11-7/8	08289	
1-1/4	#9	1-3/8	1.425	11-7/8	08290	
-	#10	1-1/2	1.550	11-7/8	08291	

A brush extension (Cat. #06282) must be used with a steel wire brush for holes drilled deeper than the listed brush length.

Fig. 2.1 Example of MPII from various product manufacturers. Do NOT use these for actual field installation. Refer to the MPII from the product that will actually be installed (Courtesy of Powers Fasteners)

G5 Adhesive Anchor Installation Instructions

RED HEAD
ADHESIVE ANCHORING SPECIALISTS

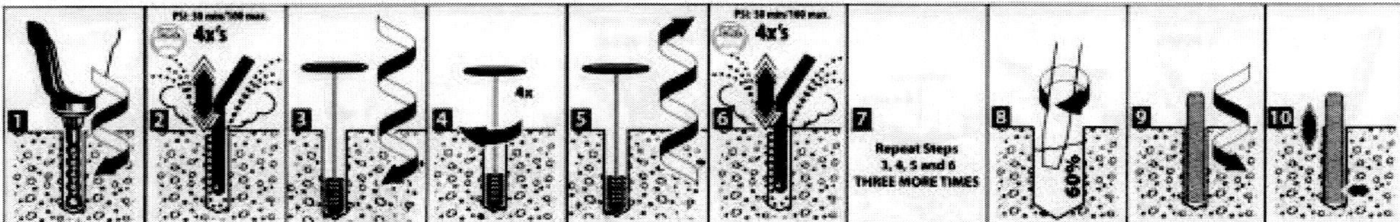

1
- Use a rotary hammer drill or pneumatic air drilling machine with a drill bit complying with ANSI B212.15.1994 tolerance standards.
- Drill depth is a minimum four times rod diameter thru maximum nine times rod diameter.
- Use a drill bit equal to the rod diameter plus 1/16" (for 3/8"ø and 1/2"ø) and 1/8" (for 5/8"ø and above).
- The 3/8"ø and 1/2"ø are acceptable for overhead, horizontal, and floor installation direction. The 5/8"ø and above permitted for horizontal and floor installation direction only.
- Per construction specification, adhere to minimum spacing, minimum edge distance, and minimum member thickness.

2
- Oscillate a clean air nozzle in and out of the hole four times, for a total of four seconds, starting at the bottom, with contaminant free compressed air exhausting hole until visually clean (i.e., no trace oils, antifreeze, et cetera).
- If hole is water filled or contains sludge/debris, the hole can be clean with pressured water for optimum cleaning method verse compressed air.

3
- Select and insert an appropriately sized Red Head wire brush from part no. SB038, SB012, SB058, SB034, SB078, SB010, SB125, or match color to anchor diameter. See chart on back page.

- Insert brush into the hole with a clockwise motion, while completing one full turn for every 1-2" forward advancement further until bottom is reached. For faster and more suitable cleaning, attach the brush to a drill. If an SDS+ drill is used, order the ESDS-38 with a 12" useable extension and the SDS+ adapter.

4
- Twist/Spin the brush four full turns at bottom of the hole.

5
- Using a clockwise motion, for every full turn of the brush, pull the brush 1-2" out of the hole.
- Air clean the dust or water clean the slurry off of brush to prevent clogging of brush.

6
- Oscillate a clean air nozzle in and out of the hole four times, for a total of four seconds, starting at the bottom with contaminant free compressed air exhausting hole until visually clean.
- Or, use pressure water to clean.

7
- **Repeat steps 3, 4, 5, and 6 (brushing and blowing) three more times** before moving to step 8.
- **Leave no dust, slurry, snow, ice or water in the hole.**

8
- Assemble Red Head (upper) and cartridge and nozzle (purchased separately).
- Place cartridge with nozzle attached into a hand or pneumatic injection tool. Dispense mixed adhesive outside of the hole until uniform color is achieved.
- Insert the nozzle tip to the far end of the hole, inject adhesive at an angle from the bottom into the hole

and leave the nozzle tip always slightly below the fill level working, in a slow circular direction, the adhesive into the side of the hole, fill slowly to prevent air entrapment, until hole is filled 60% full. After installing the anchor, gap between the rod and the concrete must be completely filled with adhesive.
- If the working time has not expired the rod can be removed from under filled holes and additional adhesive can be injected into the hole. Otherwise the installation shall be rejected. For flush application level off excess.
- For holes that contain water, keep injecting below the water and the adhesive will displace the water and prevent drawing water into the adhesive.

9
- Insert rod with a counterclockwise motion to prevent air entrapment.
- Install rod into adhesive without delay after adhesive is injected into hole during working time of the adhesive.
- Insure that the adhesive fills voids, crevices and uniformly coats rod and concrete.
- Carefully work out the air pockets with an oil, rust and scale free rod, pushing to bottom of the hole while twisting to ensure coverage.

10
- Jiggle rod slightly, immediately after installation, and do not disturb during working time.
- Allow adhesive to cure according to cure time chart on the cartridge label.

TABLE 1: SPECIFICATIONS AND DETAILS FOR INSTALLATION OF ANCHORS IN CONCRETE WITH
CPCON G5 ADHESIVE

Property		Threaded Rod Diameter (d)												
		3/8"		1/2"		5/8"		3/4"		7/8"		1"		1-1/4"	
Tensile stress area of rod (in²)	A_{se}	0.078		.142		.226		.335		.462		.606		.969	
Nominal carbide bit diameter size (in.)	d_o	7/16		9/16		3/4		7/8		1		1-1/8		1-3/8	
Effective embedment depth (in.) min/max	h_{ef}	1/2	3-3/8	2	4-1/2	2-1/2	5-5/8	3	6-3/4	3-1/2	7-7/8	4	9	5	11-1/4
Min./Max. Hole depth (in.) min/max	h_o	1-11/16	3-9/16	2-3/16	4-11/16	2-11/16	5-13/16	3-3/16	6-15/16	3-3/4	8-1/8	4-1/8	9-1/4	5-1/4	11-1/2
Minimum Slab thickness (in.)	h_{min}	4	5	4	6	4	7	5	8-1/4	8	10	8	11	10	14
Maximum Tightening Torque for pretension Clamping (ft-lb)	T_{inst}	9		16		47		90		145		170		370	

For SI: 1 inch = 25.4mm, 1 lbf = 4.45N, 1 ft-lbf = 1.356N-m, 1psi = 006895MPa

1. Adhesives must be installed in substrates at temperatures of at least 50°F to 110°F.
Installations in substrates at temperatures below 50°F or above 110°F must be conditioned to proper temperatures during working time.

2. Working time is the maximum time from the end of mixing to when the insertion of the anchor into the adhesive shall be completed.

3. Cure time is the minimum time from the end of working time to when the anchor may be torque or loaded. Anchors are to be undisturbed during the cure time.

ANCHOR INSTALLATION

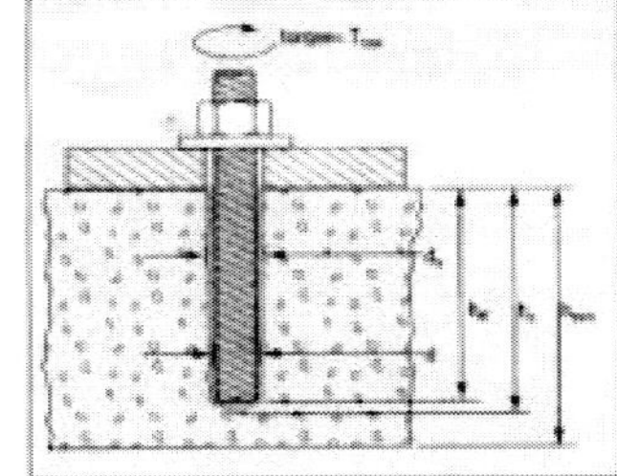

BASE MATERIAL (F°/C°)	WORKING TIME	FULL CURE TIME
70°/20°	15 minutes	24 hour

Fig. 2.2(a) Example of MPII from various product manufacturers. Do NOT use these for actual field installation. Refer to the MPII from the product that will actually be installed (Courtesy of ITW Red Head)

TABLE 2: Assembly instructions / TABLA 2: Instrucciones de ensamblado / TABLEAU 2 : Instructions d'assemblage

Assembly Instructions

1. For GS-22: Remove cap and snap off cartridge tip.
2. Thread nozzle assembly onto cartridge.
3. Release lever and pull back pistons on the Epcon Manual Injection Tool. For Pneumatic or Battery Powered Tool, retract the pistons back.
4. Insert cartridge into bracket of Injection Tool. Cartridge to be parallel with pistons during use.

Instrucciones de ensamblado

1. Para GS-22: Quite la tapa y desprenda la punta del cartucho.
2. Enrosque el conjunto de la boquilla en el cartucho.
3. Suelte la palanca y tire hacia atrás los pistones en la herramienta de inyección manual del sistema Epcon. Para una herramienta neumática o a baterías, retraiga los pistones.
4. Inserte el cartucho en el soporte de la herramienta de inyección. El cartucho debe estar paralelo a los pistones al usar la herramienta.

Instructions d'assemblage

1. Pour GS-22 : Retirer le capuchon et casser l'embout de la cartouche.
2. Visser l'ensemble de buse sur la cartouche.
3. Relâcher le levier et tirer les pistons de l'outil d'injection manuelle Epcon vers l'arrière. En ce qui concerne l'outil pneumatique ou à piles, retracter les pistons vers l'arrière.
4. Insérer la cartouche dans le support de l'outil d'injection. La cartouche doit être parallèle aux pistons durant l'usage.

TABLE 3: Brush Specifications* / TABLA 3: Especificaciones del cepillo / TABLEAU 3 : Spécifications de la brosse*

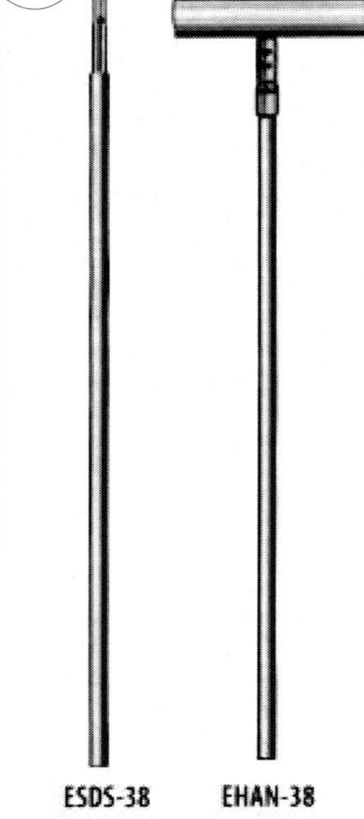

Brush Color / Color del cepillo / Couleur de la brosse	Part No. / N.° de parte. / no de pièce	Threaded Anchor Diameter (in) / Diámetro del anclaje (pulg.) / Diamètre de l'ancrage (po)	Rebar / Barra reforzada / Barre d'armature	Drill Bit Diameter (in) / Diámetro de la broca (pulg.) / Diamètre du foret (po)	Brush Diameter (in) / Diámetro del cepillo (pulg.) / Diamètre de la brosse
Grey / Gris / Grise	SB038	3/8	No. 3	7/16	5/8
Brown / Café / Brune	SB012	1/2	—	9/16	
Green / Verde / Verte	SB058	5/8	No. 5	3/4	
Yellow / Amarillo / Jaune	SB034	3/4	No. 6	7/8	1-1/4
Red / Rojo / Rouge	SB078	7/8	—	1	1-1/2
Purple / Morado / Violette	SB010	1	No. 7	1-1/8	1-5/8
Blue / Azul / Bleue	SB125	1-1/4	—	3/4	1-3/4

Part No. / N.° de parte / No de pièce	Description (Extensions come without brushes) / Descripción (las extensiones vienen sin cepillos) / Description (les rallonges sont offertes sans brosse)
ESDS-38	Steel brush 12" usable extension with SDS+ adaptor Extensión útil de 12" de cepillo de acero con adaptador SDS+ Rallonge utilisable de 12 po de brosse en acier avec adaptateur SDS+
EHAN-38	Steel brush 12" usable extension with T-handle Extensión útil de 12" de cepillo de acero con mango en T Rallonge utilisable de 12 po de brosse en acier avec poignée en T

* Proper hole cleaning using a wire brush is essential to achieve optimum performance. Brush may be used up to 50 holes depending on concrete strength. Brushes required for installation of No. 4, No. 8 rebar and larger are available with lead time.

* La limpieza adecuada del agujero usando un cepillo de alambre es esencial para lograr un desempeño óptimo. El cepillo se puede usar hasta para 50 agujeros, dependiendo de la resistencia del concreto. Cepillos necesarios para la installation the barilla No 4, No 8 y mas grandes, pueden ser obtenidos por order special.

* Un nettoyage adéquat du trou à l'aide d'une brosse métallique est essentiel à l'obtention d'une performance optimale. La brosse peut servir au nettoyage d'un nombre allant jusqu'à 50 trous suivant la résistance du béton. Les brosses recommandées pour l'installatiendes barres N° 4, N° 8 et au-dessus sont disponible avec un délai defabrication.

For faster & easier hole cleaning use the Red Head Brush in a drill

1. Use Table 3 to ensure that the proper brush is used for the hole/drill bit diameter.
2. Insert the brush into the drill and tighten.
3. Follow steps 3, 4, 5, & 6 from Table 1.

Para limpiar el agujero de manera más rápida y fácil use el cepillo Red Head Brush en un taladro

1. Use la Tabla 3 para asegurarse de que se use el cepillo adecuado para el diámetro del agujero y la broca.
2. Inserte el cepillo en el taladro y ajuste.
3. Siga los pasos 3, 4, 5 y 6 de la Tabla 1.

Pour un nettoyage de trou plus rapide et plus facile, utiliser la brosse Red Head avec une perceuse

1. Se reporter au Tableau 3 pour s'assurer d'utiliser la brosse appropriée pour le diamètre du trou/foret.
2. Insérer la brosse dans la perceuse et serrer.
3. Suivre les étapes 3, 4, 5 et 6 indiquées au Tableau 1.

ITW Red Head
800-899-7890
www.itwredhead.com
2171 Executive Drive, Suite 100
Addison, IL 60101

Fig. 2.2(b) Example of MPII from various product manufacturers. Do NOT use these for actual field installation. Refer to the MPII from the product that will actually be installed (Courtesy of ITW Red Head)

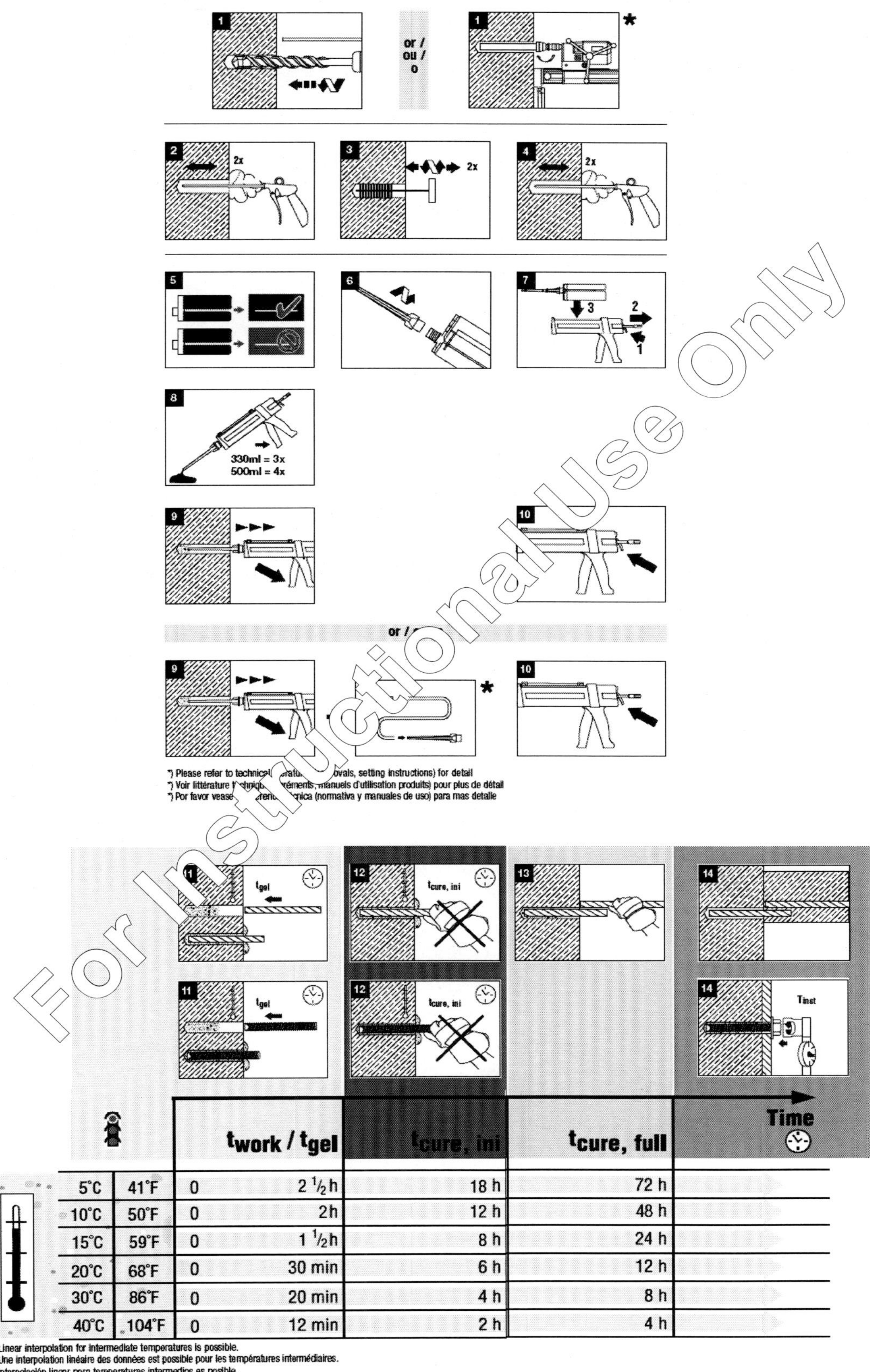

		t_{work} / t_{gel}		$t_{cure, ini}$	$t_{cure, full}$	Time
5°C	41°F	0	2 ½ h	18 h	72 h	
10°C	50°F	0	2 h	12 h	48 h	
15°C	59°F	0	1 ½ h	8 h	24 h	
20°C	68°F	0	30 min	6 h	12 h	
30°C	86°F	0	20 min	4 h	8 h	
40°C	104°F	0	12 min	2 h	4 h	

• Linear interpolation for intermediate temperatures is possible.
• Une interpolation linéaire des données est possible pour les températures intermédiaires.
• Interpolación linear para temperaturas intermedios es posible.

Fig. 2.2(b) - continued

Hilti HIT-RE 500-SD

Table 1

Ø[mm]	HAS	HIS	Rebar	HIT-RB		HIT-SZ/IP		HIT-DL	
Ø[mm]	Ø[mm]	Ø[mm]	Ø[mm]	HIT-RB	Item no.	HIT-SZ	Item no.	HIT-DL	Item no.
10	8			10	380917	–	–	–	–
12	10		8	12	336548	12	335022	12	371715
14	12	8	10	14	336549	14	335023	14	371716
16			12	16	336550	16	335024	16	371717
18	16	10	14	18	336551	18	335025	18	371718
20			16	20	336552	20	335026	20	371719
22		12	18	22	370774	22	380922	20	371719
24	20			24	380918	24	380923	20	371719
25			20	25	336553	25	335027	25	371720
28	24	16		28	380919	28	380924	25	371720
30	27			30	380920	30	380925	25	371720
32		20	25	32	336554	32	335028	22	371721
35	30		28	35	380921	35	380926	[illegible]	371721
40			32	40	382260	40	380927	[illegible]	371721

Ø[inch]	HAS	HIS	Rebar	HIT-RB		HIT-SZ/IP		HIT-DL	
Ø[inch]	Ø[inch]	Ø[inch]	Ø[inch]	HIT-RB	Item no.	HIT-IP	Item no.	HIT-DL	Item no.
7/16	3/8			7/16'	273203	–	–	–	–
1/2			#3	1/2"	273204	1/2'	274019	1/2"	38237
9/16	1/2		10 M	9/16'	273205	9/16'	274020	9/16'	38238
5/8			#4	5/8"	273207	5/8'	274021	9/16'	38238
11/16		3/8		11/16"	273209	11/16'	274022	11/16"	38239
3/4	5/8		#5 | 15 M	3/4"	273210	3/4'	274023	3/4"	38240
7/8	3/4	1/2	#6	7/8"	[illegible]	7/8'	274024	7/8"	38241
1	7/8		#7 | 20 M	1'	[illegible]	1"	274025	1'	38242
1 1/8	1	5/8	#8	1 1/8"	273214	1 1/8'	274026	1'	38242
1 1/4		3/4	25 M	1 1/4"	273216	1 1/4'	274027	1'	38242
1 3/8	1 1/4		#9	1 3/8"	273217	1 3/8'	274028	1 3/8"	38243
1 1/2			#10 | 30 M	1 1/2"	273218	1 1/2'	274029	1 3/8"	38243

Drill bits must conform to ANSI B212-1994
Les mèches de forage doivent être conformes à ANSI B212-1994.
Brocas deben cumplir con el estándar ANSI B212-1994.

Fig. 2.2(b) - continued

Setting Details of Hilti HIT-RE 500-SD with threaded rod

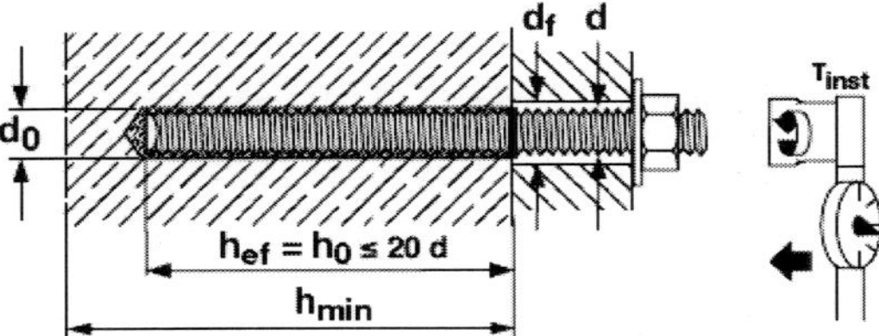

Table 2: HAS

d		d_0	h_{ef} min-max		τ_{inst}		d_f	h_{min}
[inch]	[mm]	[inch]	[inch]	[mm]	[ft-lb]	[Nm]	[inch]	[inch]
3/8	9.5	7/16	2 3/8 - 7 1/2	60 - 191	15	20	7/16	h_{ef} (30 mm)
1/2	12.7	9/16	2 3/4 - 10	70 - 254	30	41	9/16	
5/8	15.9	3/4	3 1/8 - 12 1/2	79 - 318	60	81	11/16	
3/4	19.1	7/8	3 1/2 - 15	89 - 381	100	136	13/16	$h_{ef} + 2\,d_0$
7/8	22.2	1	3 1/2 - 17 1/2	89 - 445	125	169	16	
1	25.4	1 1/8	4 - 20	102 - 508	150	20[illegible]	1 [illegible]	
1 1/4	31.8	1 3/8	5 - 25	127 - 635	200	271	1 3	
[mm]		[mm]	[mm]			[Nm]	[mm]	[mm]
M8		10	50 - 160			[illegible]	9	$h_{ef} + 30$
M10		12	60 - 200			[illegible]	12	
M12		14	70 - 240			[illegible]	14	
M16		18	80 - 320			[illegible]	18	$h_{ef} + 2\,d_0$
M20		24	80 - 400			150	22	
M24		28	96 - 480			200	26	
M27		30	108 - 540			270	30	
M30		35	120 - [illegible]			300	33	

Setting Details of Hilti HIT-RE 500-SD with HIS-N and HIS-RN Inserts

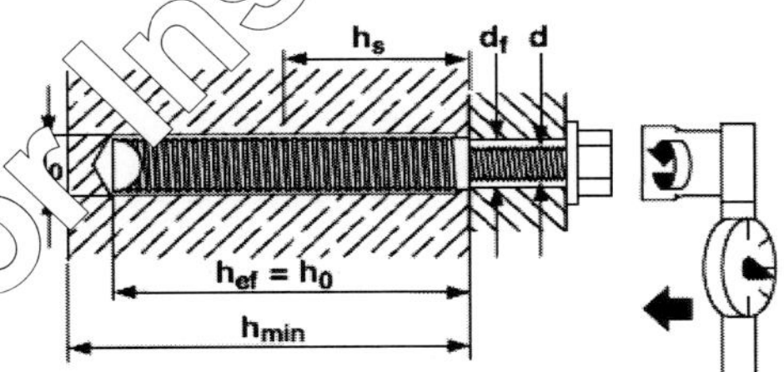

Table 3: HIS-(R)N

d		d_0	h_{ef}		τ_{inst}		d_f	h_{min}	
[inch]	[mm]	[inch]	[inch]	[mm]	[ft-lb]	[Nm]	[inch]	[inch]	[mm]
3/8	9.5	11/16	4 3/8	110	15	20	7/16	5 3/4	150
1/2	12.7	7/8	5	125	30	41	9/16	6 3/4	170
5/8	15.9	1 1/8	6 3/4	170	60	81	11/16	9	230
3/4	19.1	1 1/4	8 1/8	205	100	136	13/16	10 3/4	270
[mm]		[mm]	[mm]		[Nm]		[mm]	[mm]	
M8		14	90		10		9	120	
M10		18	110		20		12	150	
M12		22	125		40		14	170	
M16		28	170		80		18	230	
M20		32	205		150		22	270	

Fig. 2.2(b) - continued

Setting Details of Hilti HIT-RE 500-SD with reinforcement bars

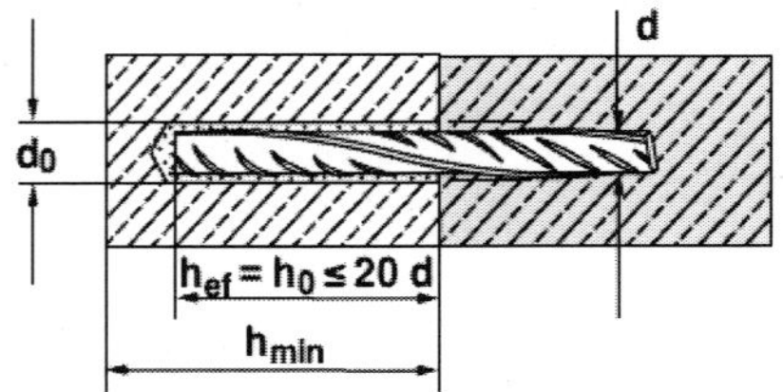

Table 4

d	d_0	h_{ef} min-max		h_{min}
US rebar	[inch]	[inch]	[mm]	[inch]
# 3	1/2	2 3/8 - 7 1/2	60 - 191	h_{ef} + 1 1/4 (30 mm)
# 4	5/8	2 3/4 - 10	70 - 254	
# 5	3/4	3 1/8 - 12 1/2	79 - 318	
# 6	7/8	3 1/2 - 15	89 - 381	h_{ef} + 2 d_0
# 7	1	3 1/2 - 17 1/2	89 - 445	
# 8	1 1/8	4 - 20	102 - 508	
# 9	1 3/8	4 1/2 - 22 1/2	114 - 572	
# 10	1 1/2	5 - 25	127 - 635	
Rebar [mm]	[mm]	[mm]		[mm]
8	12	50 - 160		h_{ef} + [illegible]
10	14	60 - 200		
12	16	70 - 240		
14	18	75 - 280		
16	20	80 - 320		
20	25	90 - 400		h_{ef} + 2 d_0
25	32	100 - 500		
28	35	112 - 560		
32	40	128 - [illegible]		
CA rebar	[inch]	[mm]		[inch]
10 M	9/16	60 - [illegible]		h_{ef} + 1 1/4 (30 mm)
15 M	3/4	[illegible] - 320		
20 M	1	[illegible]		h_{ef} + 2 d_0
25 M	1 1/4	[illegible] - 604		
30 M	1 1/2	[illegible] - 598		

Fig. 2.2(b) - continued

Hilti HIT-RE 500-SD

• **Piston plug injection - HIT-SZ/IP recommended for borehole depth > 10 inch/250 mm. Water-filled bore- holes or submerged concrete, and overhead installation the injection is only possible with aid of piston plugs.** Assemble HIT-RE-M mixer, extension(s) and appropriately sized piston plug HIT-SZ/IP. Insert piston plug to back of the borehole and inject adhesive as described in the injection method above. During injection the piston plug will be naturally extruded out of the borehole by the adhesive pressure.

11   **Insert anchor/rebar into borehole.** Mark and set anchor/rebar to the required embedment depth. Before use, verify that the anchor/rebar is dry and free of oil and other contaminants. To ease installation, anchor/rebar may be slowly twisted as they are inserted. After installing an anchor/rebar, the annular gap must be completely filled with adhesive. If the borehole is not completely filled along the embedment depth the installation should be rejected. Hilti should be contacted for further information.

   Attention! For overhead applications take special care when inserting the anchor/rebar. Excess adhesive will be forced out of the borehole - take appropriate steps to prevent it from falling onto the installer. Position the anchor/rebar and secure it from moving/falling during the curing time (e.g. wedges). Observe the gel time **"t gel"**, which varies according to temperature of base material. Minor adjustments to the anchor/rebar position may be performed during the gel time. See table.

12   **Do not disturb the anchor/rebar** once the gel time **"t gel"** has elapsed until **"t cure,ini"** has passed.

13   **Preparation work may continue for rebar applications.** Between **"t cure,ini"** and **"t cure,full"** the adhesive has a limited load bearing capacity, do not apply a torque or load on the anchor/rebar during this time.

14   **Apply load/torque after " t cure,full"** has passed, and the fixture to be attached has been positioned.

Partly used foil packs must be used up within **four weeks.** Leave the mixer attached to the foil pack manifold and store under the recommended storage conditions. If reused, attach a new mixer and discard the initial quantity of anchor adhesive as described by point 8.

Safety Instructions

For Industrial Use Only. Keep out of reach of children.

Danger: Corrosive
Harmful if inhaled or swallowed.
Can cause eye and skin burns.
Risk of serious damage to eyes.
Can cause sensitization with some individuals.
Contains quartz sand.

Precautions:
Wear suitable protection clothing, eye protection and gloves.
Do not get in eyes.
Avoid contact with the skin. Avoid inhalation of vapors.
Avoid inhalation of dusts during demolition/removal.

First Aid:
For **eye contact**, flush with water for 15 minutes while holding the eyelids apart. Seek medical attention immediately. For **skin contact**, wash immediately with soap and water. If **ingested**, drink two glasses of water and seek medical attention immediately.

Ingredient	CAS Number		
Part A: (Large side)		**Part B: (Small side)**	
Quartz sand	14808-60-7	m-xylene diamine	01477-55-0
Bisphenol A epoxy resin	25068-38-6	Aliphatic polyamine (NJ TSRN)	19136100-5014*
Bisphenol F epoxy resin	28064-14-4	Quartz sand	14808-60-7
Diglycidyl ether (NJ TSRN)	19136100-5013*	Bonding agent	65997-16-2
Alkylglycidyl ether (NJ TSRN)	19136100-5012*	Aluminum oxide	01344-28-1
Amorphous silica	67762-90-7	Amorphous silica	67762-90-7

* NJ TSNR = New Jersey Trade Secret Registry Number

In Case of Emergency, call Chem-Trec:	1-800-424-9300 (USA, P.R., Virgin Islands, Canada)
En cas d'urgence, téléphoner Chem-Trec:	1-800-424-9300 (USA, P.R., Virgin Islands, Canada)
En Caso de Emergencia, llame Chem-Trec:	001-703-527-3887 (other countries/autres pays/otros países)

Made in Germany

Net contents: 11.1 fl. oz (330 ml)/16.9 fl. oz (500 ml)   Net weight: 16.6 oz (470 g)/25.0 oz (710 g)

Warranty: Refer to standard Hilti terms and conditions of sale for warranty information.

> Failure to observe these installation instructions, use of non-Hilti anchors, poor or questionable concrete conditions, or unique applications may affect the reliability or performance of the fastenings.

Fig.2.3: Examples of MPII from various product manufacturers. Do NOT use these for actual field installation. Refer to the MPII from the product that will actually be installed (Courtesy of Hilti)

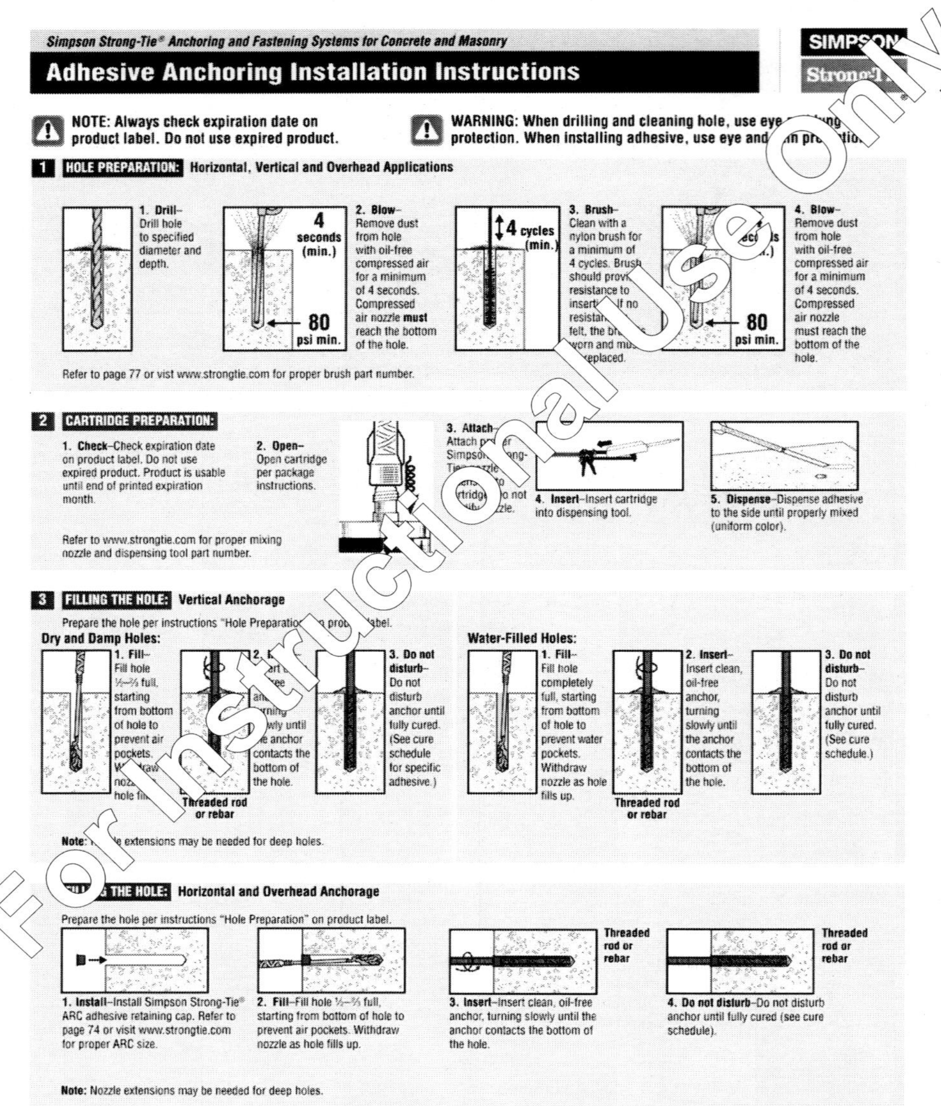

Fig. 2.4: Examples of MPII from various product manufacturers. Do NOT use these for actual field installation. Refer to the MPII from the product that will actually be installed (Courtesy of Simpson Strong-Tie)

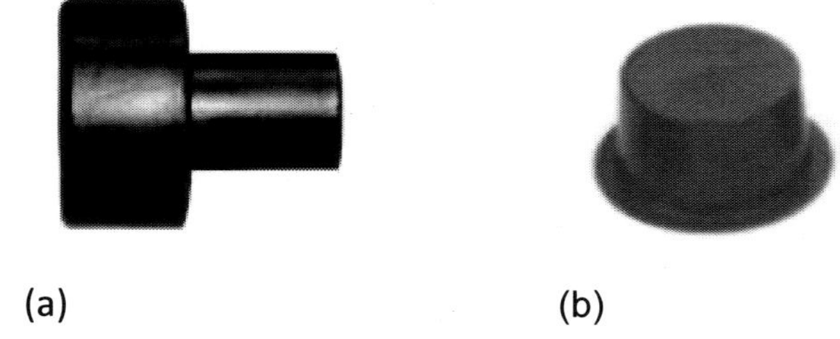

(a)   (b)

Fig. 2.5 Devices used to assist in the proper filling of horizonal or overhead applications include: (a) piston plug (Photograph courtesy of Powers Fasteners); and (b) retaining cap (Photograph courtesy of Simpson Strong-Tie)

or overhead orientations, this information is to be clearly marked on the MPII and the package labeling using symbols such as those shown in Fig. 2.7. Before installing an anchor overhead, the installer should check whether the product is suitable for overhead installation. If it is not, the installer needs to notify their supervisor.

Special instructions: Depending on the product, there may be additional instructions for proper use of piston plugs or retaining caps (Fig. 2.5). For horizontal and overhead applications, these devices are used to help place the adhesive in the drill-holes. Additionally, a retaining cap helps center the anchor. Piston plug or retaining cap sizes should be provided based on the size of the anchor to be installed.

Some MPII may provide guidance on the minimum spacing distance the anchor can be located from corners and edges and also on the minimum slab thickness to achieve desired load capacities. The minimum edge and corner distances are generally specified on the drawings or specifications for the project. Project drawings and specifications indicate the hole locations for single and multiple anchor installations. If any aspect of the MPII is unclear during review before installation, clarification must be sought before proceeding.

Material Safety Data Sheet (MSDS)

Safe handling and storage of the adhesive is very important and is generally covered in the MPII. Additional information can be found in the MSDS, which is also typically provided with each bulk purchase of anchors. Information in this workbook is provided for general background knowledge, please refer to product specific MSDS for actual procedures.

The MSDS provides necessary information, to both workers and emergency personnel, for safe use and handling of the product. MSDS include: product name, product description, emergency contact number, manufacturer, regulated ingredients/materials, physical/chemical characteristics, fire and explosion hazard data, reactivity data, health hazard data, and precautions for safe handling and use.

Many of the MSDS will be for the mixed adhesive. Some manufacturers may provide MSDS for each component and include individual component descriptions (such as color and form) prior to mixing. The product name and manufacturer contact information will be provided in case more information regarding the product is needed.

Safe handling recommendations will often include:

- Working in a well-ventilated area;
- Avoiding contact with eyes by wearing safety glasses with side shields;
- Avoiding contact with skin by wearing gloves; and
- Keeping the adhesive away from children.

Respiratory protection is typically not required. For further information review the product MSDS.

Fig. 2.6: Suggested PPE for use during adhesive anchor installation

This document does not address all safety concerns associated with the use or installation of adhesive anchors in concrete. It is the responsibility of the user to establish appropriate safety and health practices relative to the product being used and local regulation.

Personal Protective Equipment (PPE)

Personal protective equipment (PPE) is important for protection from the materials being used. PPE recommendations or requirements from the job safety specifications, employers, or the product labeling should be followed first (Fig. 2.6). In the absence of that information, the following PPE may be considered.

Suggested PPE for dust includes wearing eye protection to prevent dust from irritating the eye; respiratory protection to keep dust particles out of the nose, throat, and lungs; hearing protection to reduce the decibel exposure from hole drilling to the ear; and gloves to keep concrete dust from adhering to the skin. Adequate ventilation is also important.

At a minimum, eye protection during mixing and handling adhesive for injection is recommended to prevent any splashed or dripped adhesive from getting in the eyes. Additionally, gloves are recommended to protect the skin from contact with the adhesive.

Package Labeling

The information on the cartridges or capsules of adhesive includes pictorial selections of information from the MPII and MSDS. ACI 355.4 requires the following minimum package labeling for adhesive components:

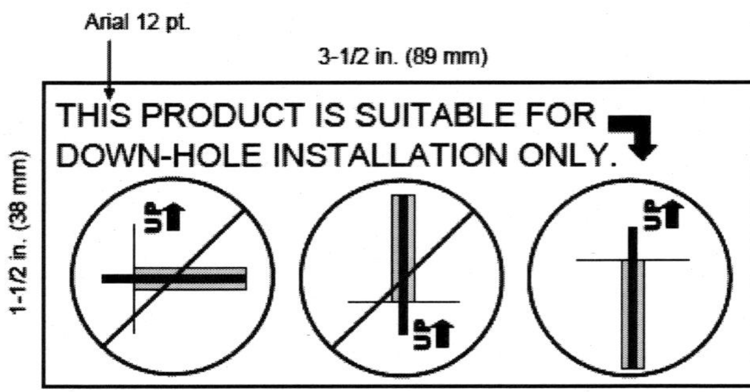

Fig. 2.7 Example of labeling required for adhesive anchor systems that are not tested for installation orientation sensitivity. Source: ACI 355.4.

- Manufacturer's name and address;
- Lot number;
- Packing date and shelf life or product expiration date;
- MPII and application information; and
- Sensitivity to installation orientation.

Figure 2.7 is an example of a label for sensitivity to installation orientation. The label shown indicates that the product is not to be used for horizontal or overhead installations. Therefore, this product should only be used for vertical down installations.

Chapter 2

1. Manufacturers provide MPII, the _________________________, that are unique for each adhesive anchor product.

    A. manufacturer's printed installation instructions
    B. marketing performance information instructions
    C. manufactured performance information instructions
    D. marketing printed installation instructions

2. MPII include directions for _________________________.

    A. drilling the hole
    B. cleaning the hole
    C. preparing the installation equipment
    D. filling the hole
    E. drill bit sizes
    F. all of the above
    G. none of the above

3. For an overhead installation, if the adhesive supplied is not qualified for overhead use by ACI 355.4 or AC308, the installer should complete the installation as instructed.

    A. True
    B. False

4. Material safety data sheet (MSDS) includes the product name, product description, emergency contact number, manufacturer, regulated ingredients/materials, physical/chemical characteristics, fire and explosion hazard data, reactivity data, health hazard data, and precautions for _________________________.

    A. injecting the adhesive
    B. expelling a dedication bead
    C. drilling the hole
    D. safe handling and use

5. Personal protective equipment (PPE) is important for installers to use during an adhesive anchor installation to protect themselves.

    A. True
    B. False

6. If an installer sees this labeling on an adhesive anchor cartridge or capsule it means this product is
________________.

    A. only suitable for overhead installations
    B. only suitable for horizontal installations
    C. only suitable for vertical down installations
    D. suitable for overhead, horizontal and vertical down installations

7. MPII from a capsule system can be used for guidance when installing a cartridge adhesive.

    A. True
    B. False

8. Gel times, found on the MPII, ________________ as the temperature of the base material increases.

    A. stay the same
    B. decrease
    C. increase
    D. double

9. The MPII includes instructions for all steps in the installation process.

    A. True
    B. False

10. Piston plugs and retaining caps are special devices for adhesive anchor installations that are
________________.

    A. overhead
    B. horizontal
    C. vertical down
    D. overhead and horizontal

Chapter 3: Drilling the Hole

An important and critical step in the anchor installation process is the hole preparation. Without a properly prepared hole, the rest of the installation process may be compromised. The drilling method to be used for anchor installation is given in the MPII and shall be strictly followed. Violation of this direction might yield fatal consequences.

For post-installed adhesive anchor installations, the MPII generally requires rotary hammer drilling as the standard drilling method. Other drilling methods such as percussion drilling or diamond core drilling might be allowed after considering the effects on load-bearing capacity or performance and especially the bond capacity. Additional hole surface treatment might also be required. In case of any deviations from the standard drilling method, the designer should always be contacted or at least informed of changes.

In the following explanations, the focus is on rotary hammer drilling because the other methods are not commonly used in adhesive anchor installation practice, although they could be used for certain production oriented projects.

The first step in hole preparation is correctly drilling the hole Fig. (3.1). Selecting the correct drill type and bit for the application can make the drilling process more efficient. A variety of drills and drill bits may be used for forming a hole for installation of an adhesive anchor. The type of drill, the type of bit, and the bit diameter to use is stated in the MPII.

Equipment: Types of Drills

There are three basic types of drills that can be used to drill holes in concrete for adhesive anchors, as shown in Fig. 3.2. They are the rotary hammer drill, the rock drill, and the core drill. The type of drill bit affects the texture of the wall in the drilled hole, which has a direct impact on the performance of the adhesive anchor system. The use of a rotary hammer drill with a carbide drill bit produces a moderately rough wall surface. Rock drill bits tend to create a hole with a very rough surface, whereas core drills create a very smooth wall surface.

Rotary Hammer Drill: A rotary hammer drill has a chuck that moves forward and backward creating a hammering motion during the drilling process. This pounding action allows the drill to pulverize the concrete more efficiently than just a rotating drill bit. Some rotary hammer drills will allow the installer to turn the hammer off. The

hammering function is usually needed for drilling into concrete.

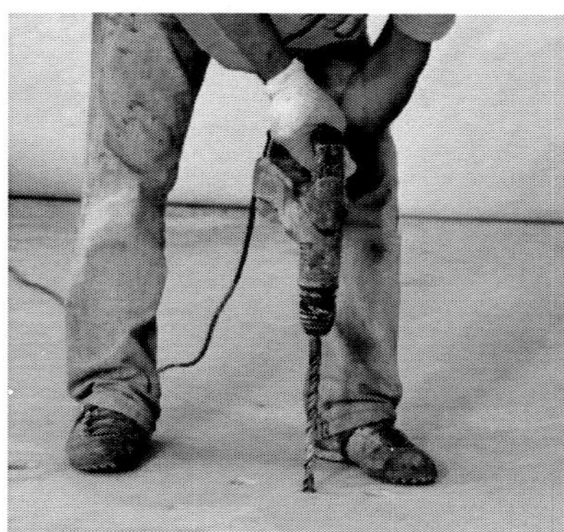

Fig. 3.1 Drilling a vertical down hole

The design of rotary hammer drills makes them suitable for drilling in concrete. Rotary hammer drills have two motions when drilling: rotation and hammering. The synchronized hammering motion breaks up and pulverizes the material, while the rotational movement distributes the hammering points around the bottom of the hole and clears the tailings and dust from the hole. Rotary hammer drills are rated based on the maximum hole diameter they can drill. Larger holes require a larger rotary hammer drill. Small holes can be drilled with a larger rotary hammer drill, but the installer must then support the additional weight of the larger drill. This can be difficult in horizontal and overhead installations.

Rock Drill: A rock drill is similar to a jackhammer. However, instead of a chisel that breaks up the concrete, a rotating bit is used to penetrate into the material. They can be powered either pneumatically or electrically. Because these tools are typically made for drilling large holes into dense, hard rock, they can drill into most concrete very quickly. These drills produce a hole with a much rougher surface than a hole drilled with a rotary hammer drill and can also damage the concrete surrounding the hole. The amount of adhesive required for a rock-drilled hole may be slightly greater than that used with a rotary hammer drilled hole because of the rougher surface. The average adhesive thickness between the anchor and the surface of the hole may be greater in a rock-drilled hole.

Core Drill: Core drills use a bit that is hollow. The bit is typically referred to as the "core barrel." Often, the open end of the barrel has diamonds embedded in the tip to increase drilling speed and core barrel life before dulling. No hammering

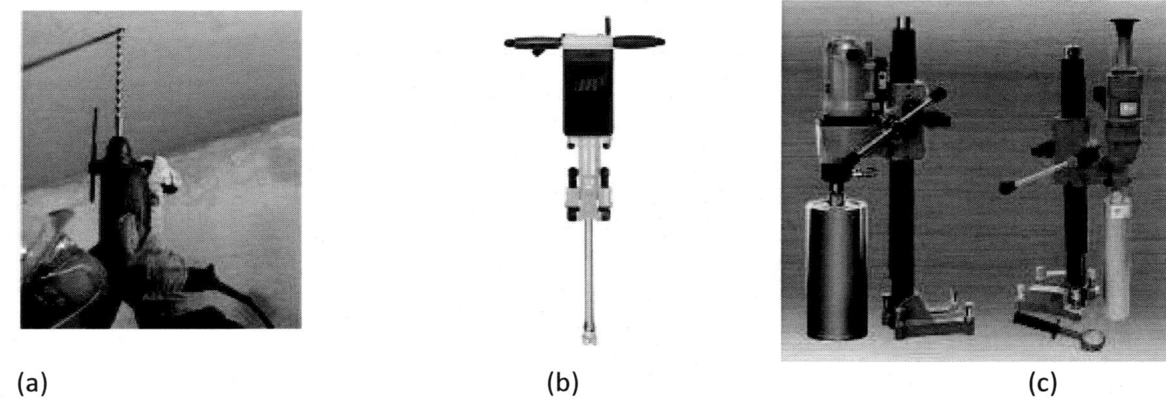

Fig. 3.2: Holes for adhesive anchor installation should be drilled using the type of drill recommended in the MPII. These drills are representative of some that may be requested: (a) rotary hammer drill (larger bits) (Courtesy of Bosch Tool Corporation); (b) rock drill (Courtesy of Ingersoll Rand); and (c) core drill (Courtesy of Ningbo Audree Industries & Trading Co., Ltd.)

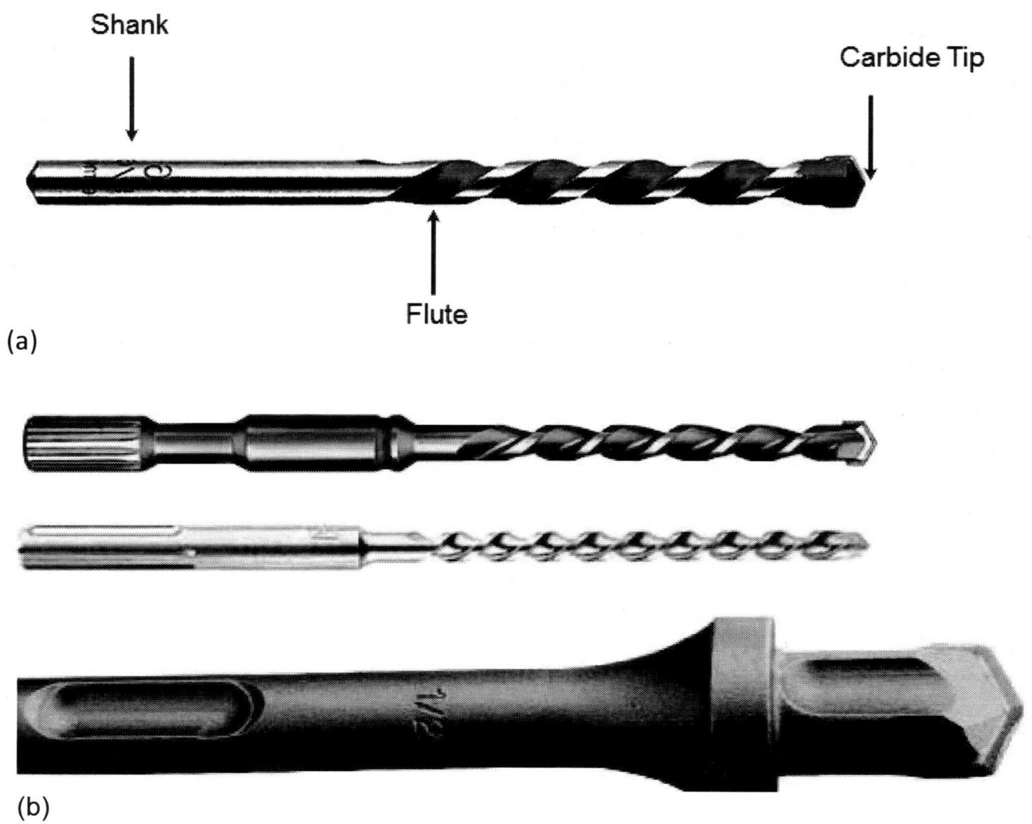

Fig. 3.3 Side view of drill bits: (a) parts of a typical drill bit (Courtesy of Bosch Tool Corporation), (b) drill bits with special shanks and shoulders (Courtesy of Milwaukee Tool and Relton Corporation)

action is used in coring, only high speed cutting action at the barrel end on the concrete creating a smooth hole surface. This is generally not desired when using adhesive anchors, as it is known to negatively impact performance. This technique can be quieter than other drilling methods but is generally more expensive and slower. Some core drilling techniques require the use of water to lubricate and cool the bit. One disadvantage of using water is that slurry is formed from the water and concrete cuttings. If the slurry is not removed while liquid, it can create a barrier between the adhesive and the concrete surface of the hole. Another disadvantage is that the hole must often

Plan View of Shank	Side View of Shank
 SDS-plus / TE-C	 Shank has two open grooves.
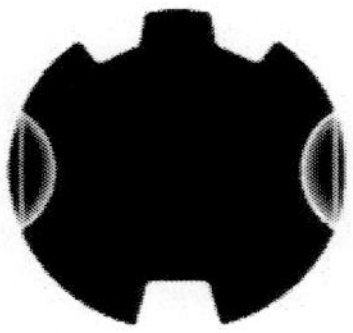 SDS-max / TE-Y	Shank has three open grooves.
 Large Spline	Shank has multiple open grooves. These are used for larger holes.

Fig. 3.4: Views of rotary hammer drill bit shanks. Special Direct System (SDS) drills are useful for heavy duty jobs such as drilling in concrete. They have three main functions: normal drilling, hammering action and chiseling (Courtesy of Bosch Tool Corporation)

be allowed to dry before the adhesive anchor can be installed unless the anchor system is specifically designed and qualified to be installed in a wet core-drilled hole. Generally, core drills are not preferred when drilling holes into concrete for the installation of adhesive anchors.

Equipment: Types of Drill Bits

Typical drill bits are shown in Fig. 3.3. The left end of the bit, shown in Fig. 3.3(a), is called the shank and is placed in the drill's chuck and fixed into position. The grooved sections in the middle and bottom portion of the drill bit are called the flutes. These are designed to direct dust and drilled material (tailings) away from the tip and out of the hole. The tip is the end of the drill bit and has first contact with the concrete. Often, the tip of a drill bit suitable for drilling in concrete will be made of a hardened material such as carbide or diamond. Drill bits typically leave 1 to 1.5-bit diameter of drill fines at bottom of the hole that can impede full insertion of the anchor if not removed (refer to Chapter 4).

Some bits are manufactured with a special shank, such as the ones shown in Fig. 3.4, designed specifically to work with a special type of drill chuck. A chuck is a specialized type of clamp, most are keyless, to hold the shank of the drill bit. The special shank and chuck system work together to secure the bit during both its rotary and hammering motions.

There are two basic styles of rotary hammer drill bits, both of which can be used for an adhesive anchor installation. A two-cutter head like the one shown in Fig. 3.5(a) has the tendency to create a triangular-shaped hole. Whereas a hole created using a four-cutter head, like the one shown in Fig. 3.5(b), will be rounder because of the greater uniformity of the head of the drill bit.

Bits used with rock drills (Fig. 3.6) often have a

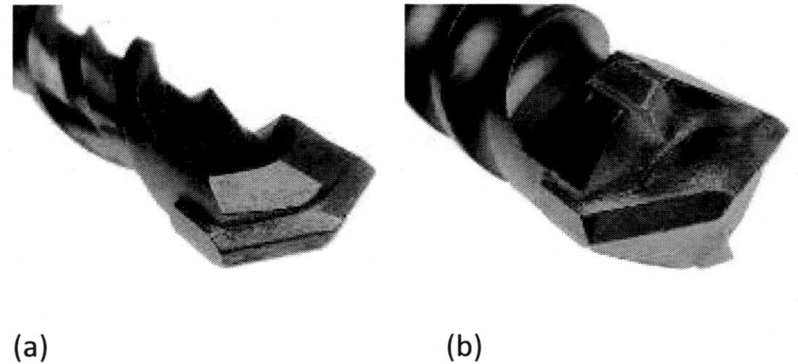

(a)   (b)

Fig. 3.5: Drill bits for use in concrete: (a) a drill bit with a two-cutter head; and (b) a drill bit with a four-cutter head (Courtesy of Bosch Tool Corporation)

Fig. 3.6 Photograph of rock drill bits (Courtesy of Tianjin Qiangli Pneumatic Tools Commerce & Trade Co., Ltd.)

Fig. 3.7 Photograph of core drill bits (or core barrels) (Courtesy of H&M Abrasives-USA)

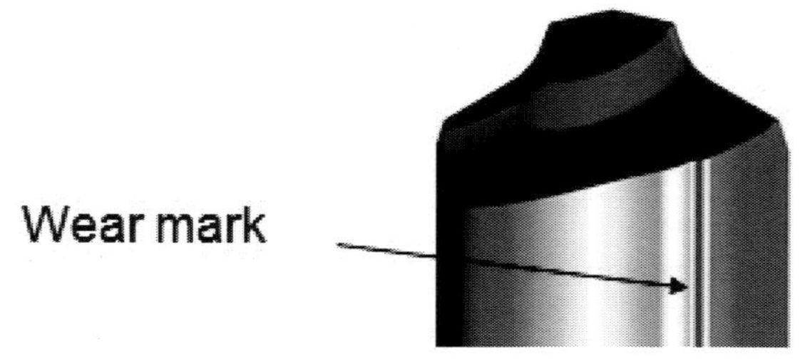

Fig. 3.8: A wear mark is an indicator that the drill bit is still suitable for use. If the mark is no longer visible, the drill bit should be discarded because the hole will be smaller than required (Courtesy of Bosch Tool Corporation)

flat cruciform head that breaks up the concrete. The shaft may or may not have auger flutes, depending on whether the bit is designed for use in drilling downward holes or upward holes. In Fig. 3.6, note the hole in the end of the rock bit. Rock drills that are air operated force pressurized air through the shank of the bit. This air blows drilling debris out of the hole and aids in cleaning.

Core drill bits (Fig. 3.7), also called core barrels, are hollow steel cylinders with a cutting edge built into the end. As mentioned previously, these bits produce a hole with very smooth walls and are generally not preferred for installing adhesive anchors. They should not be used unless specifically permitted in the MPII.

A wear mark, that may be present on the bit tip (Fig. 3.8), is an indicator used to detect worn bits. The wear mark should be checked prior to drilling to be sure the hole will have the correct size. Under most circumstances, the actual outside diameter of a drill bit is larger than the nominal drill bit diameter. For example, a nominal ½-in. diameter carbide-tipped rotary hammer drill bit will have an actual outside diameter of 0.520 to 0.530 in. when new.

The condition of the drill bit will affect the surface of the hole. The drill bit should be clean, straight, free of dirt, dust, oil, and other contaminants. The bit should not be misshapen, broken, or significantly altered from its original design. Proper care and maintenance of the drill bit will extend its useful life. A worn drill bit will not provide the correct size hole and will make drilling more difficult and time consuming. A worn drill bit may also negatively impact the performance of the adhesive anchor system being used.

Adhesive anchor manufacturers will specify the correct drill bit size in the MPII. The installer should confirm the size of the bit to be sure it meets the requirements stated in the MPII. The tolerances for carbide-tipped rotary hammer drill bit should meet the requirements of ANSI Standard B212.15, Cutting Tools - Carbide-Tipped Masonry

Nominal drill diameter	D	Tolerance range
in	in	in
1/8	0.140	+0.000 −0.006
5/32	0.171	
11/64	0.187	
3/16	0.206	+0.000 −0.008
13/64	0.221	
7/32	0.237	
15/64	0.252	
0.242	0.260	
1/4	0.268	
17/64	0.284	
9/32	0.304	
5/16	0.335	
3/8	0.398	
7/16	0.468	+0.000 −0.010
1/2	0.530	
9/16	0.592	
5/8	0.660	
11/16	0.723	
3/4	0.787	+0.000 −0.012
13/16	0.849	
27/32	0.881	
7/8	0.917	
15/16	0.980	
1	1.042	
1 1/8	1.175	+0.000 −0.015
1 3/16	1.238	
1 1/4	1.300	
1 5/16	1.367	
1 3/8	1.425	
1 7/16	1.487	
1 1/2	1.550	
1 9/16	1.608	+0.000 −0.020
1 5/8	1.675	
1 3/4	1.792	
2	2.028	

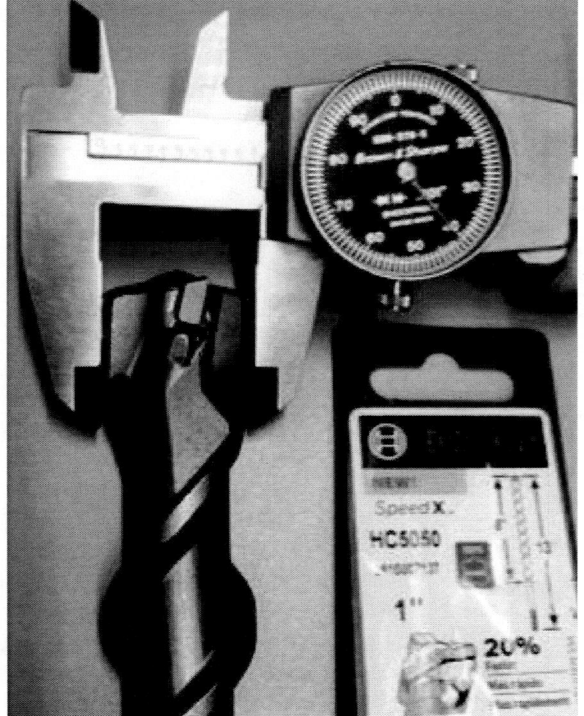

Fig. 3.9 Example using calipers to check if the drill bit tolerance is within the range stated in ANSI Standard B212.15, Table 1a[4] (Courtesy of Bosch Tool Corporation). Keep calipers perpendicular to the tip and rotate the bit in between to validate that the measurement is not on any flats

Drills & Blanks for Carbide-Tipped Masonry Drills. A micrometer or calipers can be used to check the widest part of the tip. For example, the 1-in. bit in Fig. 3.9 is specified as 1.042-in. wide when new. However, a check of ANSI B212.15 finds that the actual measured width can be +0.000 to -0.012 in. and still be within the tolerance of the standard. This means the width of the bit can be no larger than 1.042 in. + 0.000 in. = 1.042 in. and no smaller than 1.042 in. − 0.012 in. = 1.030 in. The actual measured width of the drill tip, in Fig. 3.9, is 1.041 in., which is between the two limits in Fig 3.9, so this drill bit is suitable for use.

Equipment: Selection of Drilling Machine, Drill Bit, and Drill Bit Diameter

Prior to drilling the hole, check the MPII for the correct drill type, drill bit type, drill bit size, hole depth, and alignment requirements for the product to be installed on-site. You, the installer, should familiarize yourself with the drill and bit prior to operation.

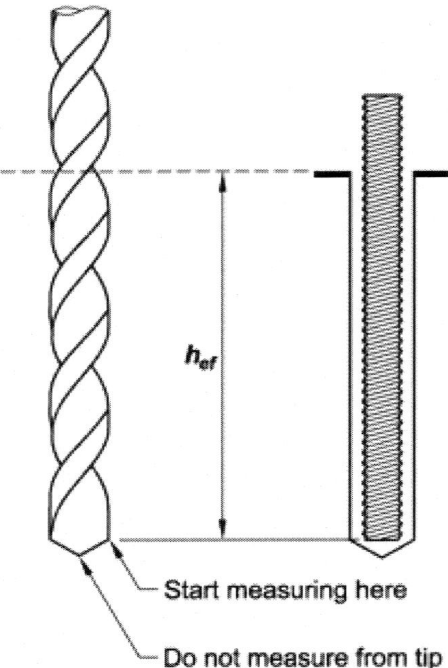

Fig. 3.10: Be sure to account for the tapered end of the drill bit when marking the length of the hole. The anchor will not be the same shape as the drill bit. Therefore the end of the anchor will not go to the tapered bottom of the hole. If this is not considered, the anchor may not get the full required embedment depth

Procedure: Drilling the Hole

Personal protective equipment (PPE) recommended and sometimes required for this operation includes safety glasses, dust mask, gloves, and hearing protection, especially for overhead usage. The bit will get hot during use, be cautious if changing bits or handling a used bit. The guidelines provided by the drill or anchor manufacturer should be read and followed.

Before drilling, check the condition of the concrete. Be sure to remove other debris and water from the concrete surface near the hole location. This step will reduce contamination of the hole with other materials. To avoid having an improperly drilled hole (for example, at the wrong location, angle, or in contact with reinforcing bars), confirm the hole location, embedment depth, edge distances, anchor diameter, and orientation by checking the construction documents. Construction documents may not specify the diameter of the hole; however, the necessary hole diameter for the anchor size to be installed can be found on the MPII.

Confirm that the equipment is in good condition and functional. To promote good contact between the bit and the drill, keep the shank of the bit free of dust and debris. Adjust the equipment according to the specifications. Verify that the drilling components selected are the tools recommended in the MPII.

Charts on the MPII will show the proper embedment depth, h_{ef}, and diameter for the hole to be drilled. Mark the depth of the hole on the drill bit (use tape flag or permanent marker) or set the depth gauge provided on the drill assembly to ensure that the hole is not shallower than required. An additional benefit of using a tape flag is that it will push dust away from the hole.

To ensure proper hole depth when using the drill bit to check the depth of the hole, be sure to account for the narrowed taper of the end of the hole (Fig. 3.10). Therefore, the hole depth when marked on the drill bit should be measured from the widest part of the tip to the required depth.

Be sure to position the drill perpendicular to the surface of the concrete or as shown on the contract documents. To ensure the drill position is maintained during drilling, secure the drill in your hand, maintain a steady drilling speed, and stop the drill at the bottom of the hole before removing the bit from the hole.

When drilling, hold the position of the drill steady so the hole is not inadvertently widened. Hole diameters are usually 10 to 25% larger than the anchor. Details are provided in the MPII. If the flutes of the drill bit get clogged during drilling, clean them to be sure they can direct dust from the hole. Wet dust creates slurry and must be cleaned periodically. During drilling, be sure to lift the drill bit up and almost out of the hole to help remove excessive dust collecting in your hole. Alternatively, use of a vacuum to remove concrete powder from the hole and dust from the surface can be considered. Either of these actions will prevent concrete fines from falling back into the hole and clogging it.

Visually inspect the debris during the drilling process. A change in color of the debris could indicate that the drill has gone into another material or that the concrete is water-saturated. If reinforcement is contacted during the drilling process stop drilling, notify your supervisor, and await further instructions.

Use a depth measuring device to ensure that the hole is drilled to the correct embedment depth. After drilling the hole, it must be cleaned. Hole cleaning will be discussed in the next chapter.

A common industry practice is to have one worker drilling the hole, another worker cleaning the hole, and yet another worker injecting the adhesive and inserting the anchor. This allows for less "change-over" time because one individual is not switching between tasks.[5] When this practice is used, it's important to coordinate the responsibilities of each person so that none of the anchor installation steps are missed.

Cited References

[4] ANSI B212.15-1994 (R2000). American National Standard for Cutting Tools. Carbide-Tipped Masonry Drills and Blanks for Carbide-Tipped Masonry Drills. American National Standards Institute, Inc. Copyright © 2002 by Cemented Carbide Producers' Association

[5] Sainz, J. "Concrete and Masonry Level 1," 2011, Robert Bosch LLC and affiliates PowerPoint presentation.

Checklist for Hole Drilling:
- Use PPE
- Verify drilling equipment
- Clear debris
- Verify hole size with anchor to be used
- Drill

Chapter 3

1. The three basic types of drills that can be used to drill holes in concrete are rotary hammer drills, rock drills, and core drills.

   A. True
   B. False

2. The hole should be drilled __________ to the concrete unless otherwise specified in the contract documents.

   A. parallel
   B. perpendicular
   C. at a 30° degree angle
   D. at a 60° degree angle

3. To ensure the proper embedment depth of the anchor, be sure to measure the required embedment depth along the length of the drill bit starting at the ____________.

   A. shank
   B. carbide tip
   C. bottom flute
   D. widest part of the tip (after the taper)

4. A wear mark is used to determine if ____________.

   A. a drill bit is still suitable for use
   B. a rotary hammer drill is broken
   C. proper PPE is needed
   D. the depth of the hole is satisfactory

5. The drill bit must be kept in good condition to provide a good bonding surface for the hole. Drill bits that are misshapen, broken, dirty or oily should ______________.

   A. be kept for a different project
   B. not be used for adhesive anchor installation
   C. be used to drill the hole
   D. be put in storage

6. Prior to drilling the hole, you have measured the width of a nominally ¼ in. drill bit tip with your calipers and the tip measures 0.264 in. Given the excerpted information from ANSI Standard B212.15, Table 1a, this drill bit is within tolerance and can be used.

Nominal drill diameter in.	D in.	Tolerance range in.
3/16	0.206	
13/64	0.221	
7/32	0.237	
15/64	0.252	+0.000
0.242	0.260	-0.008
¼	0.268	
17/64	0.284	
9/32	0.304	
5/16	0.335	
3/8	0.398	

    A. True
    B. False

7. Adhesive anchor manufacturers will specify the correct drill bit size in the ______________.

    A. MPII
    B. MSDS
    C. PPE
    D. contract drawings

8. If reinforcement is contacted during the drilling process, you should ___________.

    A. keep drilling
    B. switch drill bits
    C. stop drilling and start cleaning the hole
    D. notify your supervisor

9. The end of the drill bit that is secured in the drill by the chuck is called the _______.

    A. flute
    B. shank
    C. tailings
    D. tip

10. The tip of a drill bit is often made of hardened materials such as carbide or diamond to drill into concrete.

    A. True
    B. False

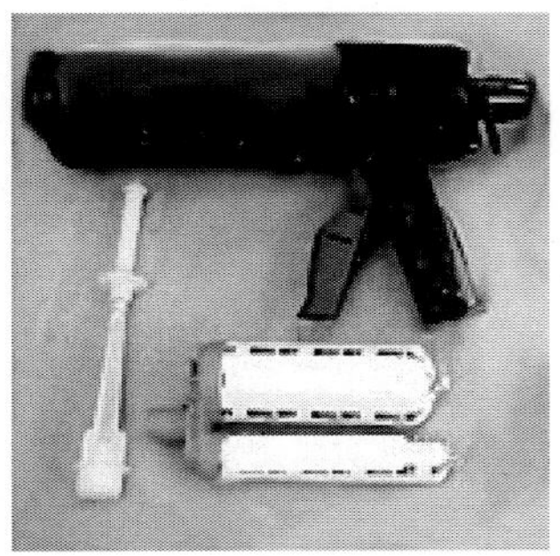 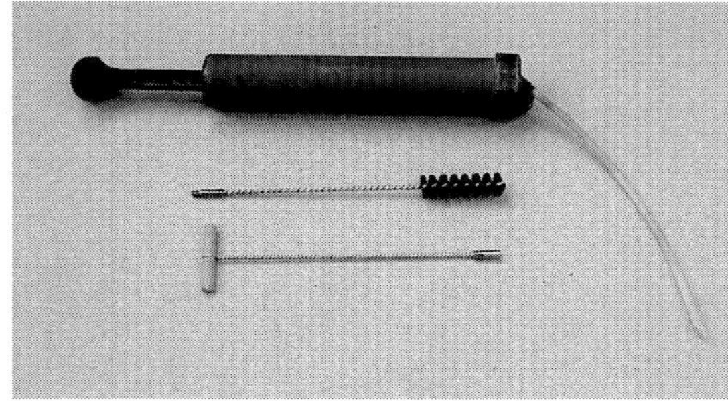

Fig. 4.1: Adhesive anchor system consisting of foil cartridge, mixing nozzle, dispensing tool, brush and hand pump (the MPII is not shown) (Courtesy of IWB, University of Stuttgart)

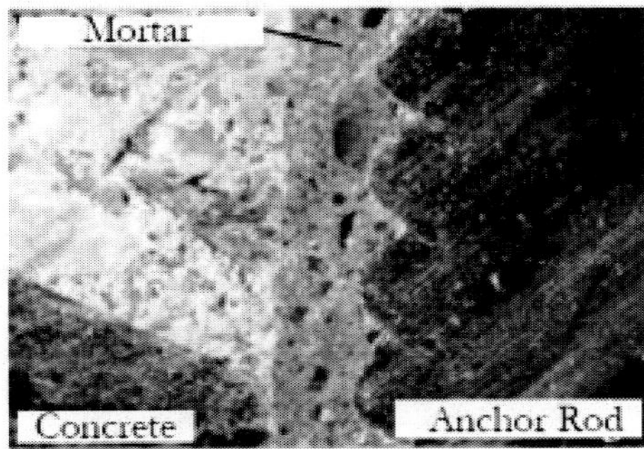 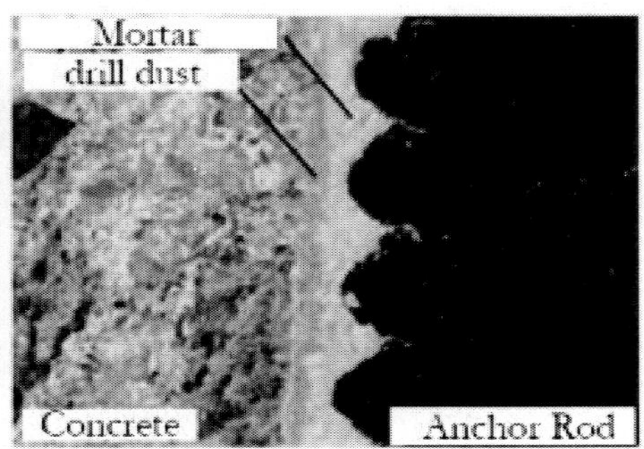

Fig. 4.2: Cut sections of installed adhesive anchors. The image on the right has a visible layer of drill dust that interferes with the bond between the epoxy mortar and the concrete. (Courtesy of IWB, University of Stuttgart)

Chapter 4: Preparing the Hole

Hole cleaning instructions vary by product and the relevant product-specific cleaning equipment is given in the MPII. The cleaning equipment for each product might vary for different diameters and embedment depths. The primary requirement is to evacuate the drill fines from the hole. An example of an injection type adhesive anchor system is given in Fig. 4.1.

It is imperative to remember, however, that the care and time spent creating a clean, roughened surface onto which the adhesive can bond will help ensure a successful installation. As shown in Fig. 4.2, a layer of dust left in the hole can prevent the adhesive from securely bonding to the concrete. The layer of dust can be compared to using flour on a baking tray to prevent cookies from sticking. Not cleaning the hole properly, as shown in Fig. 4.2, can reduce the bond strength significantly. Poor hole cleaning will reduce the performance of the overall installation (Fig. 4.3).

Cleaning procedures almost always include blowing and brushing the hole with product specific equipment. Specific products or applications may also require vacuuming or flushing with water. The number of times and the order in which these operations are performed may differ based on the MPII. For most adhesive anchor systems the brushing operation is an important step within the hole cleaning procedure.

Condition of Concrete

Provided that the adhesive was properly stored, the concrete temperature and moisture content will have a more significant impact on the bonding characteristics of the adhesive than the air temperature, to which it will have limited exposure. An inexpensive tool that can be used to check the temperature of the hole is a hand-held infrared temperature gun (Fig. 4.4). It is a highly recommended tool for an installer to use prior to injecting adhesive into the hole.

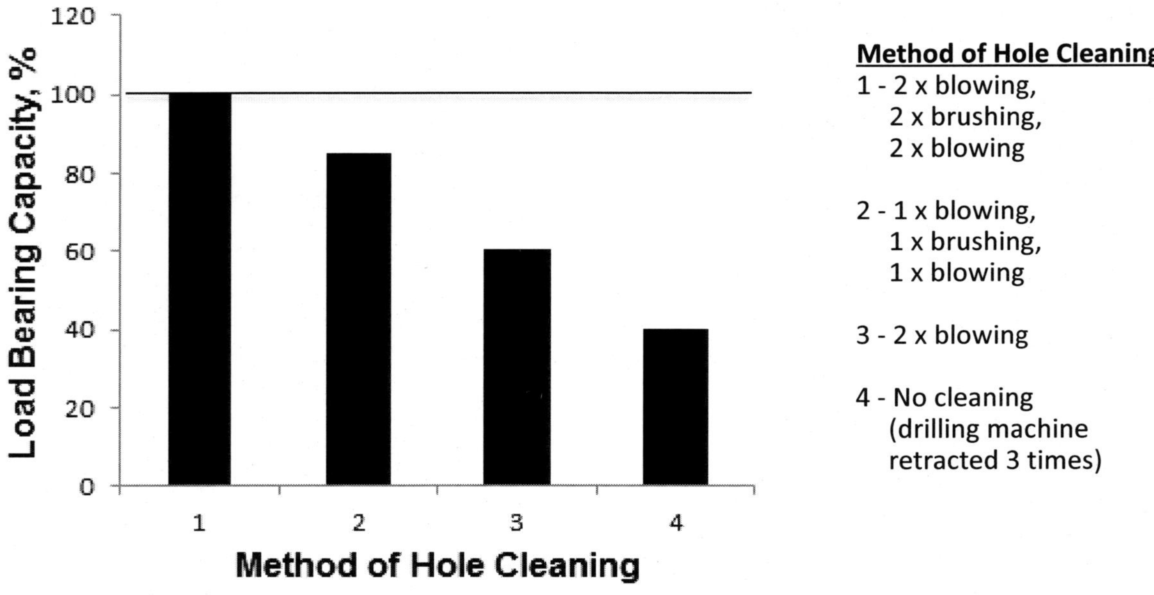

Fig. 4.3: Unclean holes can reduce bond strength up to 60% and more depending on the condition of the hole, ambient conditions (dry or humid) and the level of cleaning and the type of product. Data shown is an example valid only for dry concrete (Data for graph from Anchorage in Concrete Construction[6])

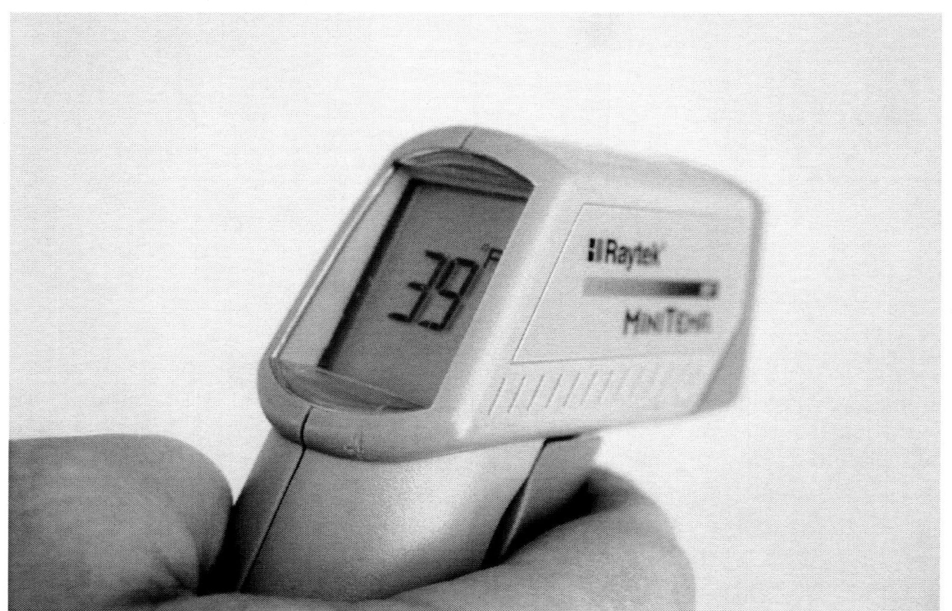

Fig. 4.4 Infrared thermometer

The concrete should be clean, dry (unless this is a wet or submerged application), and free of surface contaminants (Fig. 4.5). The ambient conditions should meet the requirements stated in the MPII. Dry concrete, per ACI 355.4-11, is concrete that is at least 21 days old and has not been exposed to water for the preceding 14 days. ACI 318-11 requires a minimum concrete age of 21 days for installation of adhesive anchors. This is to allow the concrete time to hydrate (react with water), reducing the presence of excess moisture in the concrete. This is important for adhesive anchor installations because many adhesives do not bond well in water-saturated concrete, when moisture is present, or when the adhesives are not tolerant of moist environments.

For installations where the substrate does not meet the requirements for dry concrete, check the MPII to see if the product is qualified for use in conditions where water or moisture is present.

Adhesive anchor products must be qualified per ACI 355.4-11 or ICC-ES AC308 for use in saturated, water-filled, or submerged conditions. A saturated installation is when the concrete is wet during the hole drilling and cleaning process, but there is no standing water in the hole during adhesive injection and anchor insertion. This condition can be observed during the drilling operation when the color of the drilling dust changes from grey in the beginning to a much darker grey the deeper the hole becomes. A water-filled installation is when the hole is drilled and cleaned without water present, but the anchor is installed when there is standing water in the hole. A submerged installation is when the drilling, cleaning, and installation processes are all performed with water present. Cleaning and injecting techniques may vary for the types of installations. Having a clear understanding of these situations and understanding the limitations of the product being used is the responsibility of the installer.

Weather Conditions

If the weather is dry (no rain or snow) and the concrete temperatures are within the conditions stated on the MPII, typically 30 to 100°F (-1 to 38°C), anchor installation can be performed.

If it is wet (raining or snowing) and the weather will adversely affect the condition of the hole, check with your supervisor prior to continuing the installation process. If water is present confirm that this is the intended application, check that the product to be used has been qualified for these conditions, and follow MPII cleaning instructions. If the product is not designated for use in the presence of water, do not use it under these conditions.

Equipment

The equipment to be used for the installation is product dependent and stated in the MPII. The following equipment could be mentioned:

Compressed Air: Oil-free compressed air is typically used for cleaning. This can be provided using an air compressor and wand or a hand pump. It's important to be sure that the wand (with extension if needed) can reach to the deepest part of the drilled hole. Check the wand to be sure it is free of contaminants including grease, dirt, dust, oil, and water. The MPII may state a minimum and maximum air pressure required when using an air compressor or, in the case of hand pumps, a minimum number of blows from the hand pump within a cleaning sequence.

Fig. 4.5: Image of drilled holes that need to be cleaned. The job conditions will vary on different projects. Here the installer has to work around reinforcement (Courtesy of Wiss, Janney, Elstner Associates, Inc.)

Types of Brushes: In addition to blowing air into the hole, it is necessary to brush the hole out to remove dust that has adhered to the side and bottom of the drilled hole. The brushes (Fig. 4.6 and Fig. 4.7) used for cleaning may be stiff metal or nylon. To be most effective, the brush should be clean and free of dirt, dust, oil, and other

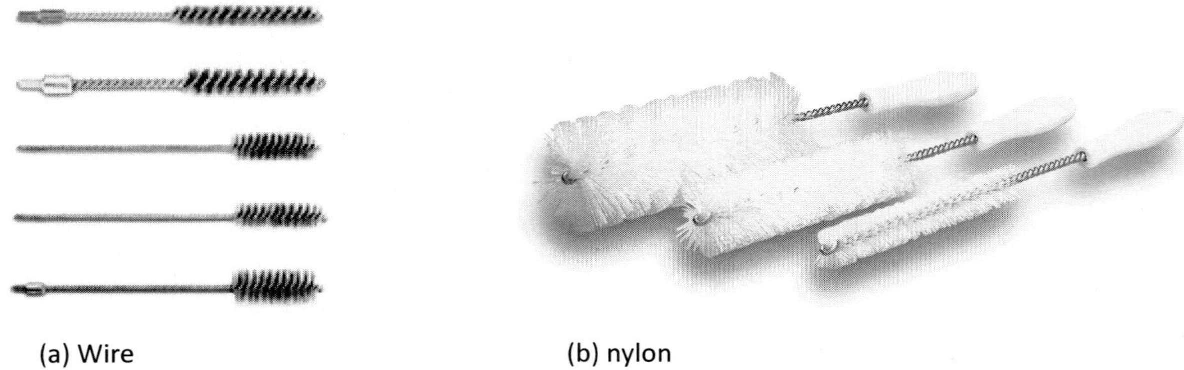

(a) Wire(b) nylon

Fig. 4.6: Examples of a variety of brushes, (a) wire (Courtesy of Sterling Hardware) and (b) nylon (Courtesy of Bunzl), suitable for specific adhesive anchor products hole cleaning

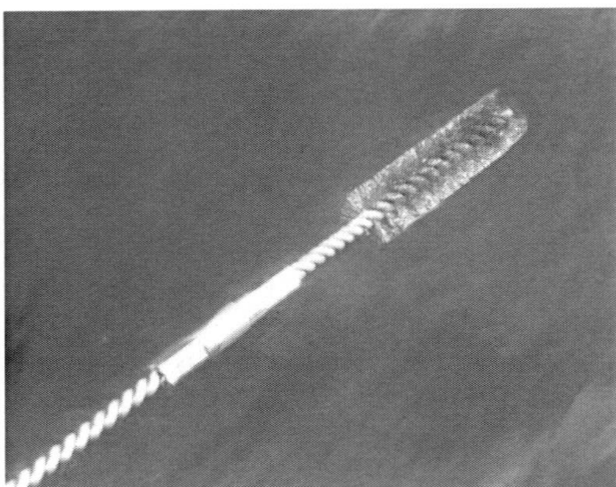

Fig. 4.7: Wire brush, example of a specific adhesive anchor product in good condition to be used for hole cleaning

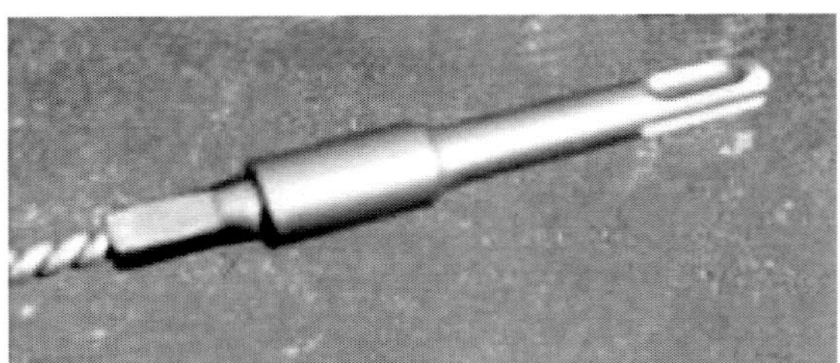

Fig. 4.8: SDS chuck, with brush attachment (brush top extends beyond left side of photograph), designed to fit into a drill

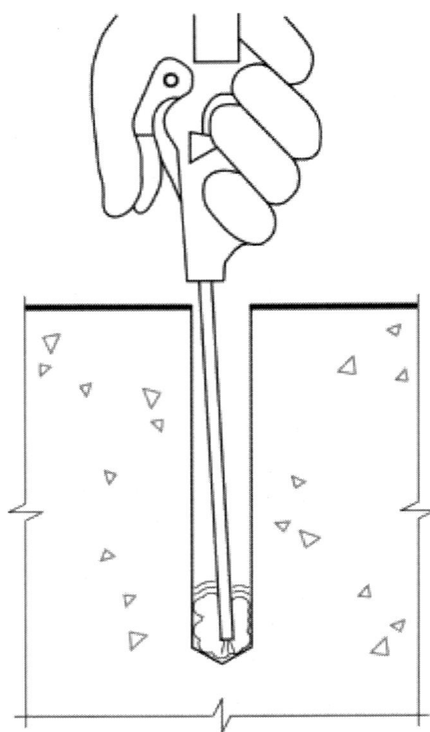

Fig. 4.9: Using compressed air to clean out a hole with a wand reaching to the deepest part of the hole

contaminants. The bristles should be full, not twisted or bent.

The size of the brush used should be as stated in the MPII. To ensure contact with the sides of the hole, the minimum brush diameter is larger than the hole diameter. If the brush does not resist insertion into the hole, it is too small or worn and should not be used. Some manufacturers provide a gauge to ensure that worn brushes are not used for hole cleaning. If brush size is important, it will be stated in the MPII.

Brushes can be manually inserted or can be powered by a drill. When using a brush with a drill, many systems have a Special Direct System (SDS) chuck to connect the brush to the drill, as shown in Fig. 4.8.

Dry Hole Cleaning Procedures

PPE recommended and sometimes required for this operation includes: safety glasses, dust mask, and gloves. Cleaning procedures for vertical down, horizontal, and overhead are the same. Care should be taken when cleaning overhead holes so debris does not drop in your face.

Determine the proper cleaning method from the MPII. In general, both cartridge and capsule adhesive anchor systems require cleaning of the hole. The following steps are based on typical instructions for dry hole cleaning.

Select the proper brush size (diameter and length) and air compression apparatus. Test the brush/drill and air compressor to be sure they are operational.

Blow out the hole using oil-free air, starting at the deepest part of the hole (add extensions if needed to reach the base of the hole) and slowly remove the wand from the hole. Perform this step as often as required by the MPII or until the air stream is free from noticeable dust.

Brush out the hole using a plunging/rotating motion as often as stated in the MPII. The brush should be inserted with a twisting motion until it makes contact with the deepest part of the hole, and then the brush should be pulled completely from the hole using a twisting motion. Some MPII may require additional full turns of the brush when it reaches the deepest part of the hole, please refer to the product's MPII for details. Brushing can be done manually or with the use of a drill. Using a drill will provide a more uniform brushing to the hole surface.

Again, *blow* out the hole using oil-free air, starting at the deepest part of the hole and removing the wand from the hole (Fig. 4.9). Perform this step as often as required by the MPII or until the air stream is free from noticeable dust.

Although the *blow-brush-blow* technique is common to most adhesive anchor products, the number of times each operation is performed will vary. If the adhesive anchor is not going to be installed immediately after the cleaning procedure, take steps to protect the hole from contamination.

In some cases, an inspector may need to examine the hole before the adhesive is injected. If the inspector questions the cleanliness of the hole because they observe dust in it, the best resolution would be to repeat the cleaning procedures in accordance with the MPII.

Wet Hole Cleaning Procedures

Follow the procedures for cleaning as outlined in the MPII.

Generally, if the hole is filled with water, submerged, or was diamond-cored (wet), the hole should be cleaned by flushing it with a water hose pressed to the deepest part of the hole, until the water runs clear. Then perform the brushing operation and repeat the water flushing. Repeat these steps as required in the MPII. Extra care to remove water-concrete particle slurries that formed during drilling should be taken. Once the slurry hardens it will form a paste layer between the hole surface and the adhesive. This layer will significantly reduce the effectiveness of the adhesive bond if not removed.

If the hole is required to be dry for installation of the adhesive anchor, use a vacuum or compressed air to remove standing water, and then use compressed air to dry the hole. The hole is still defined as 'wet' after this installation step

Checklist for Typical Dry Hole Cleaning:

- Use PPE
- Select proper hole cleaning equipment
- Blow (see MPII for number of times)
- Brush (see MPII for number of times)
- Blow (see MPII for number of times)
- Protect hole.

Checklist for Typical Wet or Submerged Hole Cleaning:

- Use PPE
- Select proper hole cleaning equipment
- Flush (see MPII for number of times)
- Brush (see MPII for number of times)
- Flush (see MPII for number of times)
- Vacuum (unless this is a submerged installation)
- Protect hole

Cited References

[6] Data for graph from: Eligehausen, R.; Mallée, R.; and Silva, J.F., *Anchorage in Concrete Construction*, Ernst & Sohn, Berlin, Germany, 2006, pp. 181-210.

Chapter 4

1. A layer of dust in the hole can ______________ adhesive bond to the concrete.

   A. improve
   B. maintain
   C. reduce
   D. enhance

2. The condition of the concrete (temperature and moisture content) will have a more significant impact on the adhesive bond strength than the air temperature.

   A. True
   B. False

3. For guidance on the cleaning techniques to be used, the installer should check the ________.

   A. contract drawings
   B. MPII
   C. MSDS
   D. PPE

4. Cleaning procedures for all adhesive anchor products will be the same.

   A. True
   B. False

5. Check the MPII to select the recommended brush size for hole cleaning. Check that the brush is not too worn by inserting it into the hole, if it _________________, it is the right size.

   A. easily fits in
   B. resists insertion
   C. does not contact the sides of the hole
   D. cannot fit inside the hole

6. The stiff metal or nylon brushes used to clean a hole should be _________________.

   A. full, straight, clean, free of dirt and other contaminants
   B. full, straight, greasy and oily
   C. worn, straight, clean free of dirt and other contaminants
   D. worn, straight, greasy and oily

7. ACI 318-11 requires a minimum concrete age of 21 days for installation of adhesive anchors.

   A. True
   B. False

8. Hole cleaning equipment for a dry installation can include _______________.

   A. brushes and a straw
   B. brushes and a vacuum
   C. brushes and gas-powered leaf blowers
   D. brushes and oil-free compressed air

9. Most MPII cleaning instructions include brushing and blowing techniques often referred to as _________, to properly remove drill dust from the hole.

   A. *blow-brush-blow*
   B. *brush-brush-blow*
   C. *blow-blow-brush*
   D. *brush-blow-brush*

10. The length of brushes and the end of an air nozzle should be checked to be sure they can reach into the deepest part of the hole.

   A. True
   B. False

Chapter 5: Cartridge Systems - Injecting the Adhesive

Adhesive anchor systems are designed as unique <u>systems</u> and have to be used as such. Using parts from multiple systems or incorrectly performing injection of the adhesive may significantly reduce the anchor load bearing capacity and might lead to an anchor failure. While many of the accessories for adhesive anchor systems may look similar and potentially interchangeable, they are absolutely not interchangeable. The accessories that are required for each system are unique to that system and vital for its correct performance. Accessories for one system should never be used with other systems. The dispensing tool, cartridge, nozzle assembly, and other accessories are designed to meter and deliver the adhesive in the proper proportions. Adhesive that is not properly proportioned will not perform as designed. Inappropriate changes in the system have significant consequences on anchor performance.

Adhesive Material Characteristics

Adhesive types: Different adhesives may be better suited for certain applications based on their curing characteristics and hardened properties. Epoxy adhesives have a long shelf life, and undergo little shrinkage during curing. Polyester adhesives have a short shelf life and can degrade under UV light and moisture. Vinyl ester adhesives have faster curing times than epoxy adhesives, but slower curing times than polyester adhesives.[7] Epoxy, polyester, and vinyl ester adhesives are exothermic (give off heat during chemical reaction of hardener and resin) during curing. Newer materials include: urethanes, methyl methacrylates, and hybrid systems of organic and inorganic bonding agents.

Temperature limitations: Storage and delivery conditions for the adhesive, specifically the temperature of the adhesive on the way from the shop to the construction site, will affect the shelf life of the product and its later performance. Each product is different; however, keeping the adhesive in a cool, dry, dark, well-ventilated storage location is best. Storage temperature ranges stated in the MPII for different anchor systems will vary. Generally, a storage temperature between 40 and 77°F (4 and 25°C) is recommended by the manufacturers. Freezing or heating of the adhesive can reduce the shelf life, as extreme temperatures can chemically alter the adhesive and its ability to react and form strong bonds which results in reduced load-bearing capacity.

Adhesive anchor installers have to recognize the temperature limitations of the adhesive products for installation. Do not install an adhesive below or above the temperature limitations provided in the MPII. The temperatures given in the MPII always refer to the concrete substrate. The temperature of the substrate governs the chemical reaction and performance of the adhesive and not the air temperature. Using an adhesive outside of the range of conditions for which it was designed and tested can cause failures and will complicate the installation process. When the products are colder than permitted by the MPII, mixing and curing will take longer or may not occur at all. When the products are warmer than permitted by the MPII, the reaction will occur rapidly and may not allow sufficient time to inject the adhesive product into the hole and install the anchor.

Gel time will vary based on the chemical properties of the adhesive and the temperature conditions of the concrete and adhesive. As shown in Table 5.1, cooler concrete temperatures extend the gel time, warmer concrete temperatures will

Table 5.1: Typical gel and curing times for an example epoxy adhesive

Temperature of Concrete (°F / °C)		Gel Time (minutes)*	Full Curing Time (hours)
40	4	180	50
50	10	120	30
60	16	75	20
70	21	30	10
80	27	25	8
90	32	15	5
100	38	12	4

Actual gel times will vary by product. Consult the MPII for the specific product being used

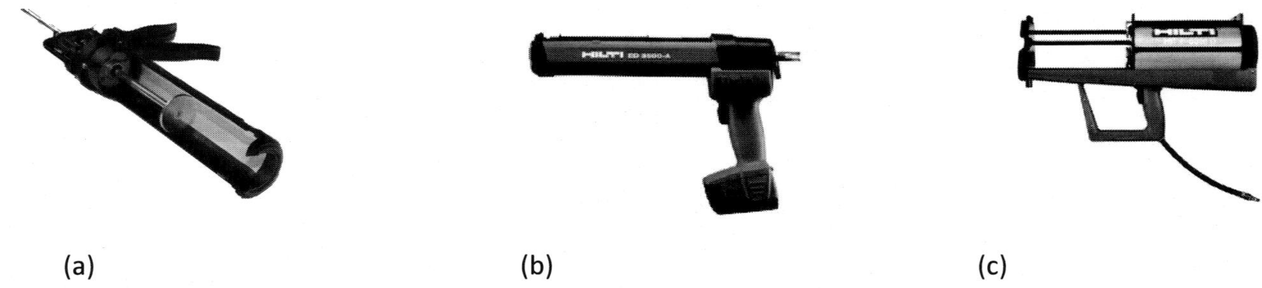

Fig. 5.1: Examples of the different types of injection tools available: (a) manual dispenser (Courtesy of ITW Red Head) (b) cordless electric dispenser (Courtesy of Hilti); and (c) pneumatic dispenser (Courtesy of Hilti)

reduce the gel time. The MPII will provide limitations on the concrete temperatures, generally in the range of 40 to 110°F (4 and 43°C). Temperature conditions of the adhesive anchor system during service life are also important. The selection of the service life temperature, however, is the responsibility of the designer.

Dispensing orientation: Dispensing characteristics of adhesives will depend on the type of product, mixing of the product, temperature of the product, and type of dispensing tool used. Some adhesives may be thicker or more fluid than others. Adhesives that can be used for horizontal to overhead applications may be thicker (more viscous) to avoid their flowing out of the hole in these orientations. Adhesives designed exclusively for use in only vertical down applications tend to be more fluid (less viscous). The movement or flow (also called rheology) of the adhesive during installation depends on a number of factors including temperature, amount and completeness of mixing, and age of the adhesive, which all affect viscosity.

Because the adhesive properties required for proper overhead installation are critical, anchor systems must be qualified per ACI 355.4 or ICC-ES AC308 for sensitivity to this installation direction. Not all adhesive anchor systems are suitable for horizontal to overhead installations. Installers should be aware of any limitations placed on the adhesive they are using. The MPII will indicate the orientations for which the adhesive is qualified.

Expiration date: Never use adhesive products that have passed their expiration date. It is important for installers to verify that each cartridge to be used has not expired. The expiration date is determined by the manufacturer. Using expired product can lead to anchor failure. Expiration dates are typically found on adhesive cartridges. Other dates and information may also be on the cartridges, such as production and packaging dates, to assist the manufacturer in quality control; therefore, it is important to make sure that the expiration date is properly identified.

Equipment

Adhesives can be designed for use in various applications, such as cold weather, warm weather, general use, sustained load, non-solid substrates (masonry), chemical resistance, and submerged and overhead installations. The equipment used for different applications may also vary. Cartridge-type adhesive systems use the following equipment: a dispensing or injection tool, adhesive cartridge, mixing nozzle, extension tubes, and brush and hand pump for cleaning purposes. A piston plug or retaining cap will also be part of the system for horizontal to overhead applications,

Injection Equipment: Manual, electric, and pneumatic injection tools are all similar in design to caulking or sealant dispensers, as shown in Fig. 5.1. A typical injection tool is designed to dispense each component of a multi-component adhesive at the appropriate ratio. The force required to dispense the adhesive is different for each system of adhesive and equipment. This should be taken into consideration, along with the number of anchors to be installed when selecting an anchor system to use on a particular project. It is good practice for the installer to test the system to assess the system's feel and rate of dispensing before installing the first anchor.

Adhesive Cartridges: The adhesive cartridge, either plastic tubes (Fig. 5.2(a)) or foil packs (Fig. 5.2(b)), which require an additional plastic holder (Fig. 5.2(c)), is placed in the injection tool. It is important to properly seat the cartridge in the tool because improper positioning of the cartridge in the dispensing tool may result in damage to the cartridge or incorrect pressure on the components. Both may result in poorly mixed adhesive. The cartridge will have a sealing cap or twist-off-tip on the cartridge to which the mixing nozzle attaches, sometimes referred to as the manifold. This cap must be removed prior to use. Initial pressure supplied from the injection tool should break any remaining seals in the cartridges and allow the two adhesive components to flow through the

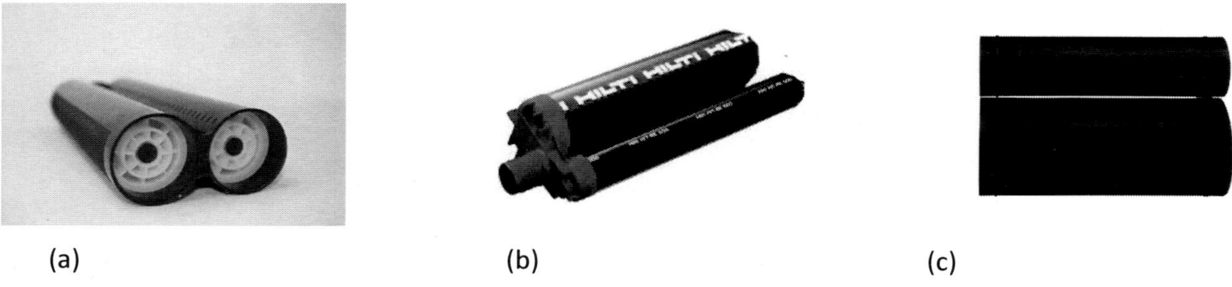

(a)   (b)   (c)

Fig. 5.2: Examples of adhesive packaging for cartridge adhesive systems: (a) adhesive cartridge in plastic tubes (Photograph courtesy of ITW Red Head); (b) adhesive cartridge in foil pack (Courtesy of Hilti); and (c): foil pack holder (Courtesy of Hilti).

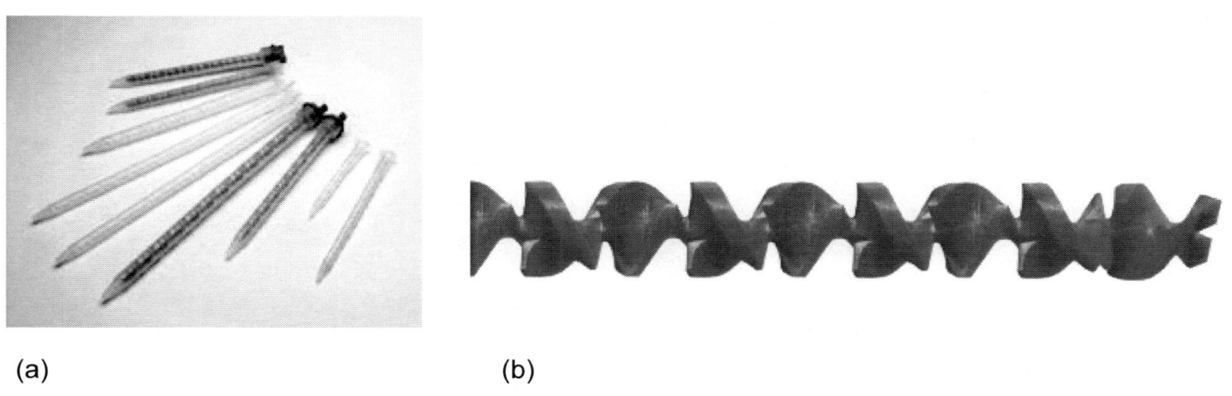

(a)   (b)

Fig. 5.3: Mixing nozzles come in different sizes and lengths: (a) Examples of typical mixing nozzles (Courtesy of ITW Devcon); and (b) static mixing element designed to mix the two components as they pass through the nozzle (Courtesy of Powers Fasteners)

manifold to the mixing nozzle.

Mixing Nozzles: The mixing nozzle is attached by the installer to the manifold of the cartridge. Mixing nozzles, shown in Fig. 5.3(a), must not be cut or altered in any way or other mixing nozzles substituted for the system's nozzle. The proper mixing nozzle is designed to ensure proper mixing of specific adhesive components. Inside the mixing nozzle is a static mixing element (Fig. 5.3(b)). The adhesive components are mixed as the material travels through this element. It is not permitted to remove the mixing element from the mixing nozzle.

The mixing nozzles are designed with a retaining nut on one end to attach the nozzle to the cartridge manifold. Many retaining nuts are molded to the nozzle. If needed, an extension tube or an additional piece of plastic tubing can be attached to the dispensing end of the nozzle. This extension tube is used to reach the deepest part of the hole if the end of the mixing nozzle cannot. The extension tube is the only part within the adhesive anchor system that is allowed to be modified and therefore can be cut to any length

needed. Additional comments on extension tubes with respect to dispensing an initial amount of adhesive to ensure the proper mix ratio are made later.

Piston Plugs: A piston plug (Fig. 5.4) is a device used to help uniformly fill overhead holes with adhesive from the deepest part of the hole. It is best to attach the piston plug to the end of a flexible extension tube, rather than the end of the rigid mixing nozzle. The extension tube and piston plug are inserted into the hole until the piston plug touches the deepest part of the hole. The piston plug allows adhesive to flow through it to produce pressure on the dispensing side of the piston plug (the deepest part of the hole). The pressure of the flowing adhesive will gently force the piston plug toward the installer (out of the hole) ensuring optimal filling. Maintaining a slight bend (slack) in the flexible extension tube allows the piston plug to be pushed out of the hole. A common mistake of installers is to pull on the extension tubing/piston plug assembly while injecting adhesive. This action may create voids in the adhesive (Fig. 5.5). A good practice for installers is to use both hands to hold

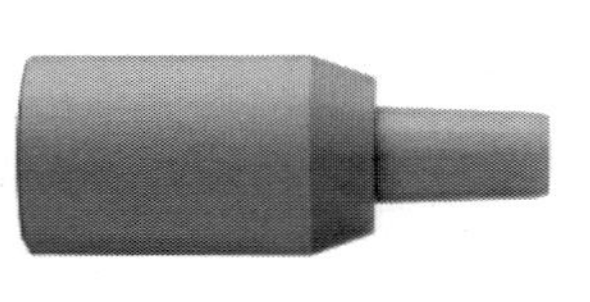  

Fig. 5.4 Piston plugs (Courtesy of Hilti, ITW Red Head, and Powers Fasteners)

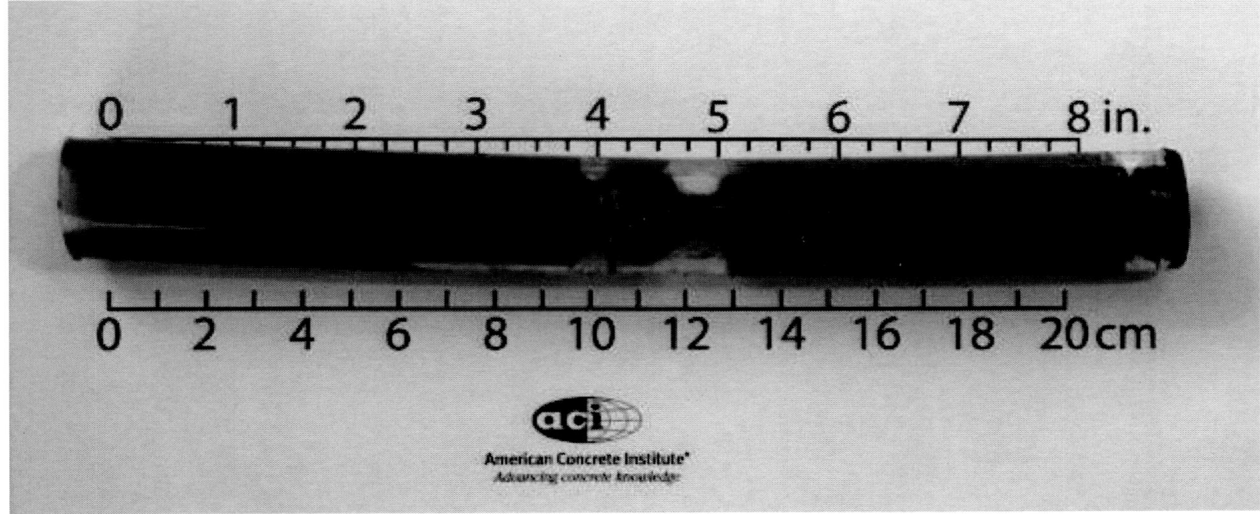

Fig. 5.5: Example of poorly filled tube where the piston plug was pulled out too quickly and not allowed to be extruded by the power of the adhesive. This pulling creates voids in the adhesive. The top of the tube is positioned on the right

Fig. 5.6 Retaining caps (Courtesy of Simpson Strong-Tie)

the injection tool after inserting the piston plug and inject while maintaining slack in the line; this will prevent the installer from being able to pull on the extension tubing and piston plug.

Retaining Caps: A retaining cap (Fig. 5.6) is another device designed to aid in horizontal and overhead applications. The cap is placed at the opening of the hole and is used to help retain the adhesive inside the hole. The mixing nozzle or extension tube is inserted through cross cuts in the retaining cap to fill the hole with adhesive.

Procedure: Injecting the Adhesive

The safety guidelines provided by the adhesive anchor system manufacturer must be read and followed. Personal Protective Equipment (PPE) often recommended for this operation includes safety glasses and gloves with the additional caution to be sure the work area is properly ventilated.

Prior to inserting the adhesive, it is critical to verify the condition of the hole, the concrete temperature, and that all the required equipment for the installation is available and operational. The overall condition of the adhesive should also be checked. Be sure to look at the expiration date and check that the foil pack or cartridge has not been punctured or previously opened. Do not leave the adhesive exposed to direct sunlight.

The hole should be clean (re-clean the hole per the applicable MPII if the condition of the hole is unknown), dry (if applicable), and moderately rough on the wall surfaces with the surrounding area free of water and debris. To verify the hole depth, a clean anchor, pre-marked with the embedment depth (either by marker or tape), can be inserted into the hole. Proper hole diameter may be determined by inserting a properly-sized drill bit for the anchor to be installed (per the MPII) and verifying that it fits into the deepest part of the hole with minimal resistance. If there is any reason to believe the hole diameter is incorrect, check the diameter of a select number of holes and refer to the product MPII to check for conformance.

The ambient conditions, temperature and humidity near the hole, should be within the allowable range of the adhesive. Typical working temperatures for an adhesive are between 40 to 100°F (4 to 38°C). Typical temperature requirements for the concrete are between 40 to 110 °F (4 to 43°C). In any case, check the MPII for specific temperature requirements. The temperature of the adhesive and the concrete will affect the gel time and cure time of the adhesive. Anticipation of the gel time and planning the installation pattern are important when there are multiple holes to fill. In any case, the gel time must be sufficient to ensure proper insertion of the anchor rod. Warmer temperatures will speed up curing, that is, reduce gel time, and may require replacing the mixing nozzle before the cartridge is emptied. The adhesive temperature should not go below or exceed the storage temperatures stated in the MPII.

You should always review the MPII and MSDS for the adhesive system you are using. Don't assume all the MPII and MSDS from all manufacturers are the same. Each product is unique and has specific guidelines to follow. Assembly of the components is manufacturer specific. Be sure to review the provided MPII. Ensure that the nozzle is properly assembled to the adhesive cartridge manifold, and check that the cartridge is positioned correctly in the dispensing tool. Confirm that the nozzle (or piston plug) can reach the deepest part of the hole. If necessary, use a flexible extension tube to reach the deepest part of the hole. The tubing should attach securely to the nozzle. Duct tape can be used to secure the tubing to the nozzle.

Visually check that the adhesive dispensed from the nozzle is consistent (uniform color) before injecting it into the hole. The required procedure is given in the product's MPII: A small amount of adhesive, sometimes referred to as a dedication bead (Fig. 5.7), is typically discharged into a box or other disposable container to confirm uniform mixing of the adhesive through the mixing nozzle. It is not acceptable to have non-uniform adhesive (observed by non-uniform color or streaks) when

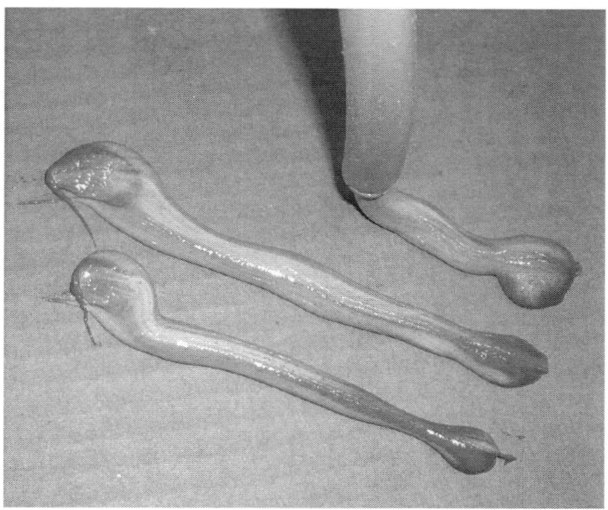

(a)

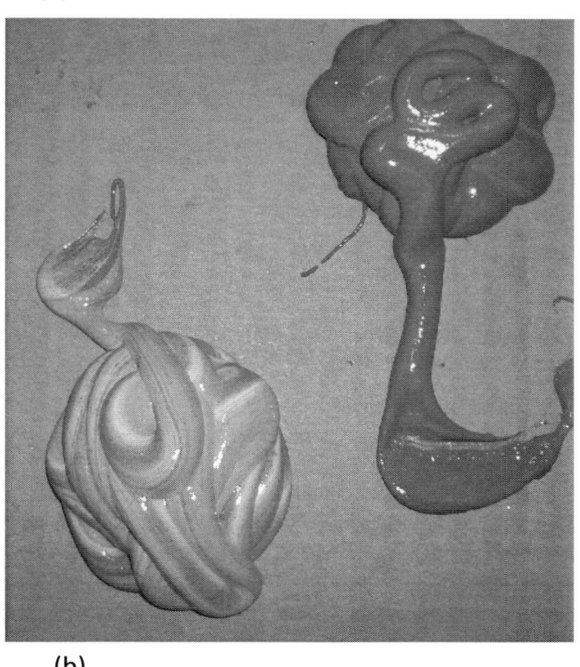

(b)

Fig. 5.7: The MPII for each product will indicate how much adhesive should be discarded at the start of a new cartridge. This material, also known as the (a) dedication bead, is discarded to ensure that the product used in the installation is well-mixed (no marbling). The photograph on the right (b) shows unmixed adhesive and well-mixed adhesive

filling the hole. If the adhesive is not properly mixed, it will not be able to achieve its full performance characteristics.

To help determine that the hole has been filled with sufficient adhesive, a mark can be placed on the extension tube or nozzle. Prior to filling the hole, place a mark from the tip of the tube or piston plug down one third the depth of the hole (Fig. 5.8, 5.9, and 5.10). During the adhesive injection, the

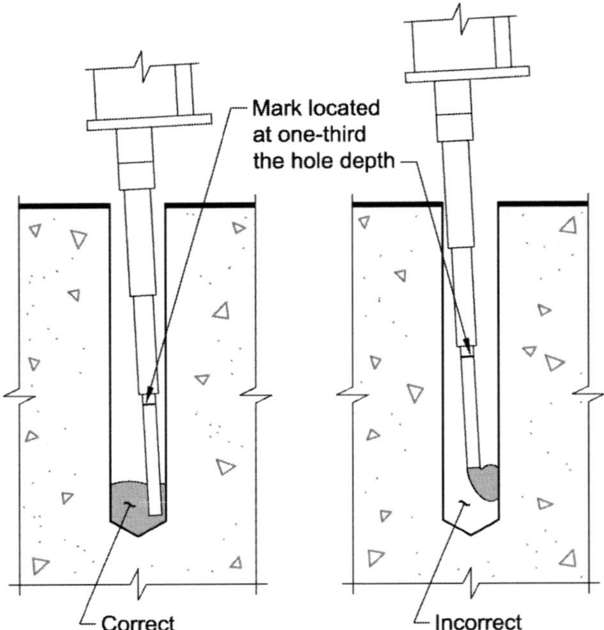

Fig. 5.8: Using a cartridge adhesive system to fill an anchor hole. The dispensing end of the nozzle should reach to the deepest part of the hole to ensure complete filling. Mark located one-third the hole depth on the nozzle helps to determine proper fill level

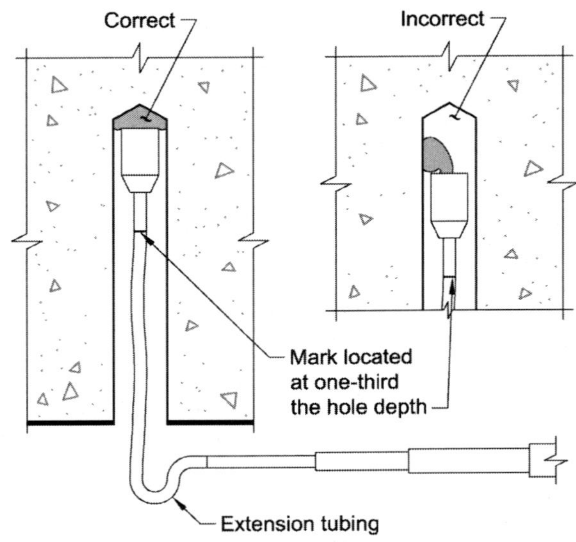

Fig. 5.9: Using a cartridge adhesive system with flexible tubing and a piston plug to fill an anchor hole. The dispensing end of the piston plug should reach to the deepest part of the hole to ensure complete filling

hole has been filled two-thirds full when this mark is even with the concrete surface.

Vertical Down: To inject the adhesive, start at the deepest part of the hole (Fig. 5.8) and use a smooth, continuous motion to dispense the adhesive. Removing the nozzle too quickly during filling may cause unwanted air voids in the adhesive. Keep the tip of the nozzle submerged in the deposited adhesive throughout the filling process. Fill the hole approximately two thirds full with adhesive. Slowly remove the nozzle from the hole.

If there are no more holes to fill and the cartridge is not empty, the mixing nozzle may be left on the cartridge as a cap. The adhesive may be used at a later date by replacing the old mixing nozzle with a new one. When reusing a previously opened cartridge, be sure to use a new mixing nozzle. Also, dispense initial adhesive (dedication bead) prior to filling the hole. When the cartridge is empty properly dispose of the nozzle and spent cartridge in accordance with applicable regulations. Typically, if adhesive components are still in the cartridge and it is not desired to save them for later use, they may be injected into a receptacle and allowed to harden for disposal. Caution must be used when disposing of adhesives that generate heat as they can melt or burn the disposal receptacle or materials with which it comes in contact. Each time a new cartridge is used, a new mixing nozzle must also be used.

Horizontal or Overhead – piston plugs: Some systems used in horizontal or overhead applications include a piston plug designed to facilitate complete filling of the hole. If this system is used, the MPII will include information on the proper size piston plug to use. The piston plug will be slightly smaller in diameter than the drilled hole. When filling a hole, the piston plug is inserted to the deepest part of the hole (Fig. 5.9). As the adhesive fills the hole, the piston plug is pushed toward the opening of the hole. Filling should be completed in a continuous, smooth motion as noted previously.

Even if it's not required, it may be beneficial to connect the piston plug to the mixing nozzle with a length of flexible extension tubing. This allows the installer to maintain a slight bend (slack) in the extension tubing (Fig. 5.9). The slack allows the piston plug to be pushed out of the hole by the pressure of the adhesive being injected into the hole. Without the flexible tubing, the piston plug is attached directly to the rigid nozzle, and it may be more difficult for the installer to allow the pressure from the adhesive to push the piston plug out. It is important to allow the piston plug to be pushed out of the hole by the adhesive throughout the injection procedure. Voids will be formed in the hole if the nozzle is pulled out too quickly.

Overhead – retaining caps: A retaining cap is used in horizontal to overhead applications to help retain the adhesive in the hole as well as center

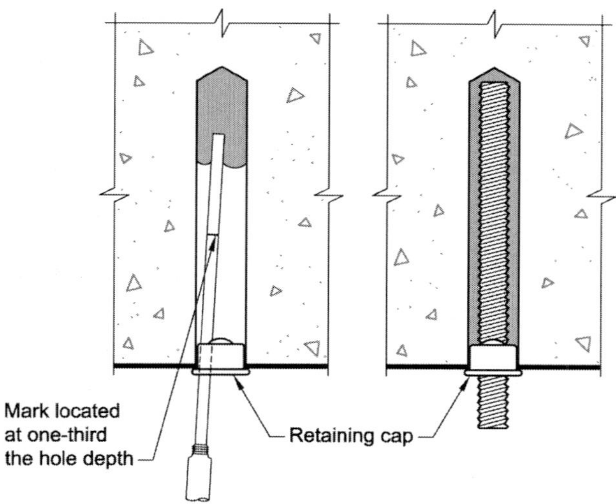

Fig. 5.10: Some adhesive anchor systems use a retaining cap in overhead installations to retain the adhesive until it cures. An "X" cut in the cap allows the mixing nozzle or nozzle extension and the anchor to pass through the cap

(a)

(b)

Fig. 5.11: Completed vertical down installations: (a) excess adhesive that has extruded from the hole should be wiped off before it hardens if it will make it difficult to place an attachment on this anchor; and (b) adhesive that is level with the surface

the anchor in the hole until the adhesive sets (Fig. 5.10). The cap is positioned at the opening of the hole and has a cross cut in the center to allow the nozzle to pass through for filling and later for when the anchor is installed. Be careful not to widen the cut or shift the cap with the mixing nozzle during filling. The MPII will include information on the size of the retaining cap to use.

When using a retaining cap system, it may be helpful to mark a point on the nozzle one third the hole depth from the end with masking tape. This will ensure that the installer knows when they reach the desired level. Simply marking the nozzle with a marker may not work because the mark may become covered with adhesive during filling. The use of the tape acts as "speed bump" because you feel it hit the retaining cap as you pull out the nozzle.

It is helpful to "wet-out" the hole by moving the end of the nozzle or tubing in a circular motion at the deepest part of the hole. This helps secure the initial adhesive material to the bottom and sides of the hole and prepares it to accept additional material. This method can help eliminate or reduce voids in this critical part of the installation.

Verification of adhesive level in the hole is done visually. It is important to be sure the hole is at least 2/3 full. It is not necessary to fill the hole completely. When the anchor is inserted it will displace the adhesive and spread adhesive onto the concrete, anchor, or both. Excess adhesive on the concrete surface or anchor can create difficulty when attaching a base plate, washer, nut, or other attachment, but this may be cleaned up immediately after anchor installation (Fig. 5.11). Not having enough adhesive in the hole after inserting the anchor must not be corrected by injecting additional adhesive into the remaining void around the top of the anchor. If the adhesive does not reach the level of the concrete surface after installing the anchor, remove the anchor and consult a supervisor. Typically in such situations, the adhesive is allowed to fully cure, the hole is re-drilled, and the installation process repeated properly.

Special Case / Wet Conditions

Check with your supervisor if the adhesive is suitable for wet condition applications.

When injecting adhesive into a hole with free-standing water present, use precautions to prevent trapping water between the adhesive and the concrete. Also take care to prevent trapping pockets of water within the adhesive. Techniques to prevent air entrapment described in this chapter should also be used to prevent water entrapment.

Cited References

[7] Wisser, Erich K.R.; Kunz, J.; and Geiss, P., "Behavior and Design of Adhesive Anchors Under Tensile Load." Proceedings of the 23rd Annual Meeting of the Adhesion Society, Myrtle Beach, 2000.

Checklist for Injecting the Adhesive:

- Use PPE
- Review MPII
- Check hole depth and anchor length
- Check hole condition
- Verify and assemble injection equipment
- Mark the extension tube or nozzle
- Discard initial material (dedication bead)
- Fill hole according to the MPII, rule of thumb: 2/3 full

Chapter 5

1. All adhesives for adhesive anchor installation have the same fresh and hardened properties.

    A. True
    B. False

2. Storage of adhesives should be done in accordance with the MPII; generally, they should be stored in a ___________, well-ventilated storage location.

    A. warm, wet, bright
    B. warm, dry, dark
    C. cool, wet, bright
    D. cool, dry, dark

3. Before using an adhesive, check the _______________ date to ensure that the product performance will be at its fullest capacity.

    A. expiration
    B. production
    C. packaging
    D. shipping

4. The ___________________ is where the two adhesive components from a cartridge flow into the mixing nozzle.

    A. dispenser
    B. manifold
    C. retaining nut
    D. static mixing element

5. The only part in an adhesive anchor system that is allowed to be modified is the ___________.

    A. cartridge adhesive
    B. extension tube
    C. mixing nozzle
    D. piston plug

6. When using a piston plug in an overhead or horizontal installation, it is important to pull firmly on the plug and extension tubing to ensure a void-free hole.

    A. True
    B. False

7. Confirm that the mixing nozzle, extension tube or piston plug can reach _________ the hole.

    A. two thirds into the depth of
    B. the closest part of
    C. the deepest part of
    D. the middle part of

8. At the start of a new cartridge, check the MPII to determine how much initial material, also called a ____________, should be discarded to ensure that the adhesive used is properly mixed.

    A. dedication bead
    B. tacky tear drop
    C. waterfall effect
    D. void

9. When injecting a hole with adhesive, be sure to place the end of the mixing nozzle or piston plug _________________ of the hole.

    A. at the deepest part
    B. at the middle part
    C. at the closest part
    D. two thirds into the depth

10. When injecting the adhesive, a hole should be ___________ full to ensure that there is sufficient adhesive to properly complete the installation.

    A. one-quarter
    B. one-half
    C. two thirds
    D. completely

Chapter 6: Inserting the Anchor (Cartridge Systems)

After injecting the adhesive into the hole, the clean anchor can be installed according to the product specific MPII. Care should be taken to properly insert and position the anchor in a manner that does not entrap air. If the MPII provides methods for supporting and protecting the anchor, these must be followed.

Anchors

Anchors include continuously threaded rods (Fig. 6.1) and deformed reinforcing bars (Fig. 6.2). Most anchors are fabricated from carbon steel of a variety of strengths, but stainless steel anchors are also available for better environmental corrosion resistance. The most common anchor type is a continuously threaded rod where the diameter ranges from 3/8 in. to 1 1/4 in. Other special anchors (larger diameters can be used with special review), such as an internally threaded insert that will accept a threaded bolt (Fig. 6.3), are also available. Deformed reinforcing bars may also be used as anchors. Details are found in the product specific MPII.

Anchors (threaded rod, reinforcing bar, inserts) may have no coating or they may be epoxy-coated or galvanized. If the anchor is cut from threaded rods or reinforcing bars, cleaning to remove dirt, oil, grease, or other contaminants is required. Be sure to use cleaning methods and agents that do not harm the coating. For loose dirt and other dry debris, a wire brush or clean cloth may be used to remove the contamination. For oil, grease, or other sticky contaminants, a solvent or detergent may be required. Be sure to remove not only the contaminants, but the solvent or detergent as well.

The MPII may provide guidance on suitable cleaning methods and agents. It is important to remove any contaminants from the anchor surface prior to insertion. Failure to do so will prevent the adhesive from forming complete bond to the anchor and could cause failure of the anchor system. Similar to Fig. 4.1, where lack of hole cleaning left a layer of dust between the concrete and the adhesive interface that caused a reduction in the adhesive-concrete bond, oil, grease, dirt, or other contaminants present on the anchor will reduce the bond between the adhesive and the anchor.

Procedure

Inserting the Anchor — PPE recommended and sometimes required for this operation includes safety glasses and gloves. The guidelines provided

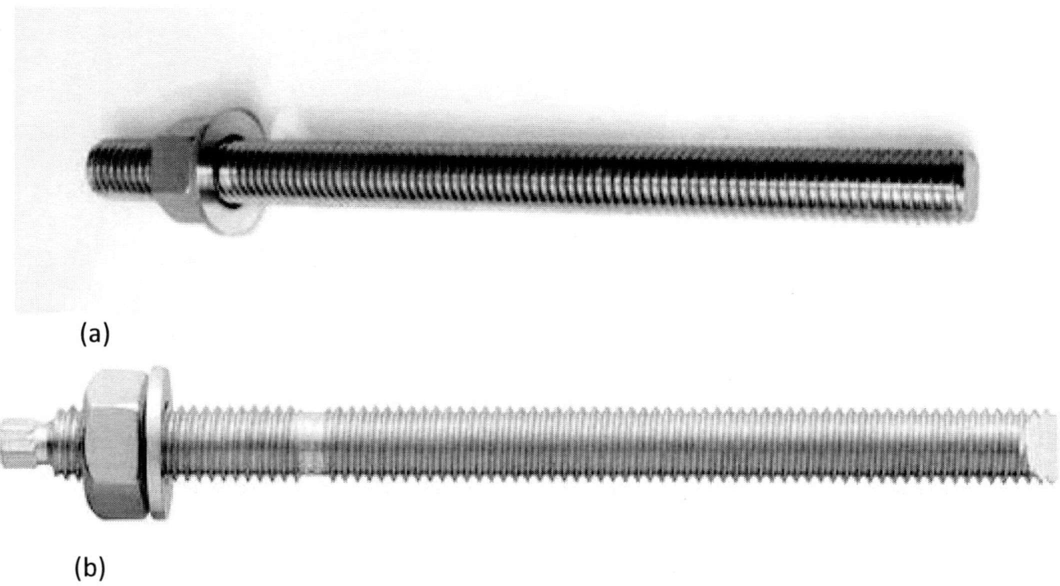

(a)

(b)

Fig. 6.1: Examples of threaded rod anchors: (a) a length of continuously threaded rod (Courtesy of ITW Red Head); and (b) a continuously threaded anchor rod with embedment mark and tapered end optimized for adhesive anchor applications (Courtesy of IWB, University of Stuttgart)

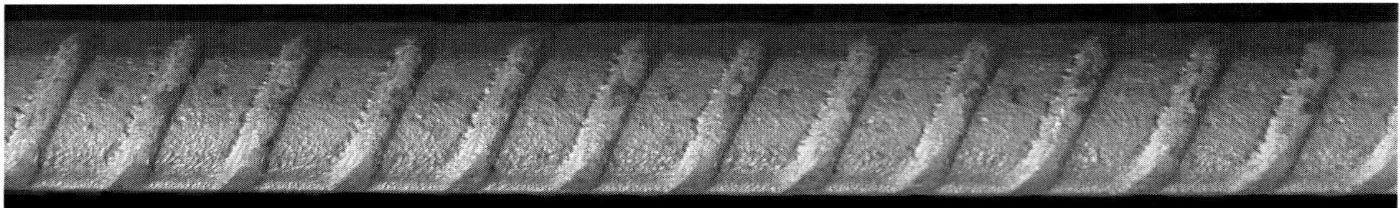

Fig. 6.2: Example of deformed reinforcing bar that is used as an anchor

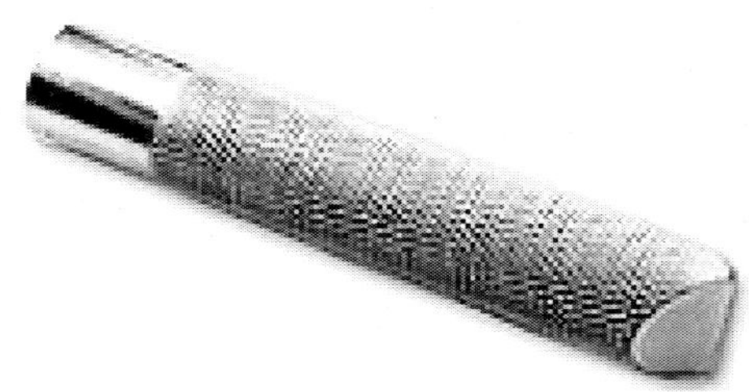

Fig. 6.3: Example of an internally threaded anchor designed to accept a threaded bolt or rod (Courtesy of Alma Bolt Company & Prime Fasteners)

Fig. 6.4 Example using tape to mark the embedment depth and protect the threads from getting filled with adhesive during installation

by the MPII should be read and followed.

Prior to injecting adhesive, check to be sure that there is enough adhesive left in the cartridge to sufficiently fill the hole and that an acceptable anchor is available. The anchor should be inspected to ensure that it is dry and that any dirt, oil, grease, rust, loose mill scale, or other contaminants have been removed. In the case of cut anchors, it is good practice to mark the embedment depth on the anchor with a permanent marker or removable tape (Fig. 6.4). The benefit of using tape is that it also protects the threads from filling with adhesive

during installation. If marking with tape, make sure that the bottom of the tape is positioned such that it does not interfere with the adhesive-anchor-concrete bond. Note that the tape must not be removed from the anchor before the adhesive is fully cured. Anchors should be checked for uniform threading, required length, correct diameter, and warping. The use of bent or damaged anchors should be avoided. Severely damaged threads may affect the strength of the anchor or prevent the nut from threading onto the anchor. If a nut cannot be threaded onto one end of the anchor due to damaged threads, but the other end is usable, it may be permissible to install the anchor with the damaged threads in the hole. A bent anchor may be difficult or impossible to fully insert into the hole. The hole should be checked to confirm that it is at least two thirds full of adhesive prior to inserting the anchor. This is most often done visually by estimating the distance from the top of the hole to the level of the adhesive or by marking the injection nozzle (Fig. 5.8, Fig. 5.9, and Fig. 5.10).

When inserting the anchor, rotate it slowly to ensure positive distribution of the adhesive. Check the MPII to see if there is a preferred direction to rotate the anchor while inserting. If no preferred direction is stated in the MPII, it has been observed that when inserting a threaded rod, a counter-clockwise rotation may help retain more adhesive in the hole. Be sure that the anchor is fully seated in the deepest part of the hole so that the required embedment depth is achieved. Confirm the embedment depth by checking the mark placed on the anchor. Align the anchor perpendicular to the substrate (unless otherwise indicated in the project documents). Typical industry standard for anchor alignment is ± 3 degrees from perpendicular (Fig. 6.5). Deviations beyond 3 degrees may introduce a bending load that will change the performance of the anchor and may make if difficult or impossible to place base plates or similar attachments.

Check that the adhesive has pushed all the way up the hole to the surface of the concrete. Confirm that the gap between the hole wall and the anchor is completely filled with adhesive (Fig. 6.6). If there is an excess of adhesive on the concrete, remove it prior to adhesive hardening using a trowel or a piece

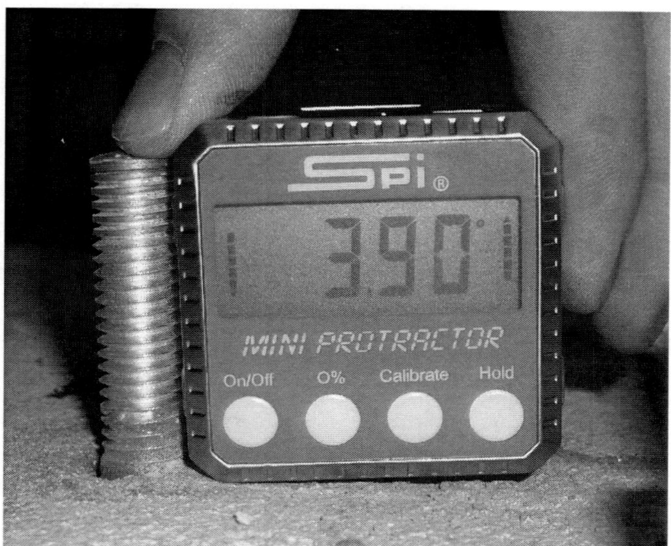

Fig. 6.5: Checking perpendicularity or plumbness of the anchor after installation. Alignment should be a maximum of ± 3 degrees from perpendicular

Fig. 6.6: Examples of vertically down installed anchors. The anchor on the left is under filled and represents a poor installation. This anchor should not be loaded. During injection of the adhesive, the installer should have placed more material in the hole. The anchor on the right is an example of a properly inserted anchor with the adhesive filled to the top of the hole

of hard cardboard being careful to leave adhesive in the hole level with the surface of the concrete without disturbing the anchor. When inserting the anchor, be sure to protect the exposed portion of the anchor from the adhesive as this may adversely impact the attachment by covering the threads or bar.

Anchors must be inserted during the gel time of the adhesive. Minor adjustments to the position and slight jiggling of the anchor to fully seat the anchor during the gel time are allowed. After the gel time has passed, protect the anchor from further disturbance until the adhesive is fully cured. It is not permitted to completely remove and re-insert an anchor during the gel time. If an anchor is completely removed, it should be discarded. If an anchor is removed, the adhesive remaining in the hole should be allowed to fully cure, after which the hole may be re-drilled, cleaned, injected with adhesive, and a new anchor installed.

Under filled holes: If there is not enough adhesive in the hole, the installation must be repeated. Remove the anchor, allow the adhesive to fully cure, re-drill the hole, and repeat the installation procedures.

Overfilled holes: If there is too much adhesive in the hole, take care to not allow the adhesive to adhere to the exposed portion of the anchor.

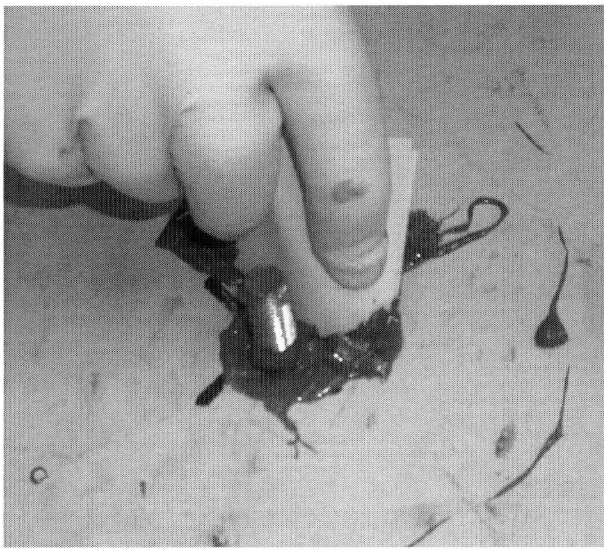

Fig. 6.7: Removing fresh adhesive without disturbing the anchor

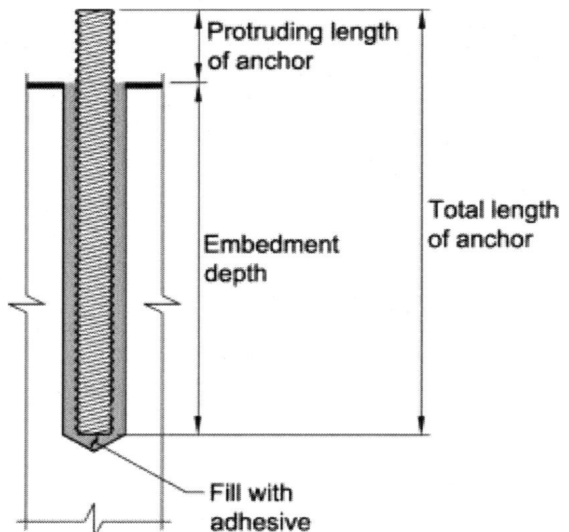

Fig. 6.8: The total length of the anchor minus the length of the protruding portion is the embedment depth.

Remove as much adhesive from around the hole opening as necessary to make the adhesive level with the concrete surface (Fig. 6.6). Excessive adhesive can prevent the proper attachment of a washer and nut or other assembly to the surface of the concrete. If the excess adhesive was not removed prior to curing (Fig. 6.7), it can be carefully chipped away after it has fully cured.

Retaining caps: When inserting an anchor through a retaining cap, be sure to not shift the cap out of position. The cap should assist in centering the anchor in the hole as well as retain the adhesive. The proper amount of adhesive is present when a minor amount is forced out of the cap when the anchor is inserted. Retaining caps are not designed to support large amounts of weight from anchors in overhead orientations, but can with special consideration and designer review. Use additional support and protection as recommended in the MPII.

Support and Protection: In an overhead installation, provide support to the anchor, using wedges, caps, screens, or other restraint methods, until the adhesive has fully cured. No additional devices, aside from the anchor, wedges (if acceptable per the MPII), retaining cap (if applicable), and screen (if applicable) should be inserted into the adhesive without review or per the MPII.

For all installations, do not disturb the anchor until the adhesive has fully cured for the time listed in the MPII. Nuts on the ends of the anchors should not be tightened until the full curing time has passed. Verify the final embedment depth based on the length of the protruding anchor. If the total length of the anchor is known and the protruding length is measured then the embedment depth is the difference (Fig. 6.8). Also, visually check the perpendicularity of the anchor. The anchor should not be altered after installation. After curing, a fixture can then be attached to the anchor.

Chapter 6

1. This photograph is an example of a _______________ that is commonly used as an adhesive anchor. These are available with diameters ranging from 3/8 in. to 1-1/4 in.

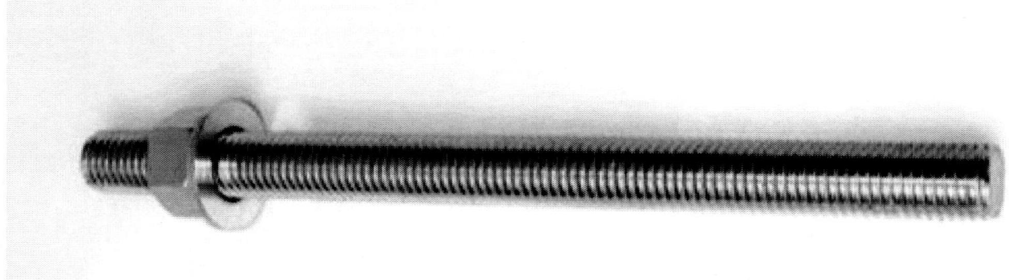

   A. continuously threaded rod
   B. deformed reinforcing bar
   C. internally threaded insert
   D. J-bolt

2. This is an example of an under-filled hole. Because the adhesive level is too low, this anchor should be _________________.

   A. placed into service
   B. removed and the adhesive allowed to harden so the hole can be re-drilled and the anchor replaced
   C. injected with more adhesive
   D. removed, and the hole should immediately be filled with additional adhesive, and a new anchor installed

3. Before use, an anchor should be inspected to be sure it is _________________.

   A. dry, clean, straight, and marked with the embedment depth
   B. dry, dirty, straight, and marked with the embedment depth
   C. wet, clean, straight, and marked with the embedment depth
   D. wet, dirty, straight, and marked with the embedment depth

4. A threaded anchor should be checked to be sure that the threads are intact. If a nut cannot be threaded on one end of the anchor, check if the other end is usable.

   A. True
   B. False

5. As the anchor is inserted into the adhesive _________________ to ensure even distribution of the adhesive on the anchor and to retain adhesive in the hole.

   A. jiggle it rapidly
   B. press it quickly
   C. rotate it slowly
   D. spin it swiftly

6. Anchor alignment should be within ___ degrees from perpendicular to the surface of the concrete.

   A. 3
   B. 6
   C. 9
   D. 12

7. If excessive amounts of adhesive, which can prevent the proper attachment of a washer and nut or other assembly to the surface of the concrete, are pushed out of the hole onto the surface of the concrete by insertion of the anchor _________________.

   A. it can be removed with a trowel while the adhesive is fresh
   B. it can be carefully chipped away when it has hardened
   C. it can be removed with a piece of cardboard while the adhesive is fresh
   D. all of the above

8. You have measured an anchor that is 6.5 in. After installation there is 2.5 in. protruding from the concrete. What is the final embedment depth of the installed anchor?

   A. 2.5 in.
   B. 4.0 in.
   C. 6.5 in.
   D. 9.0 in.

9. You are placing a base plate over a four anchor installation. You notice that three of the four anchors are the same height and the fourth anchor is an inch shorter. What may have occurred?

   A. fourth anchor hole was drilled too shallow
   B. fourth anchor hole was drilled too deep
   C. fourth anchor was cut too short
   D. fourth anchor was cut too long
   E. B and C
   F. none of the above

10. For supporting an overhead anchor following installation use ________________.

    A.   a wedge
    B.   an extra anchor
    C.   a mixing nozzle
    D.   an empty cartridge

Chapter 7: Capsule Adhesive Anchor Systems

In some applications, capsule adhesive anchor systems can be an alternative to cartridge adhesive anchor systems. Each type of adhesive system has advantages that may be important for a particular project: vertical down, horizontal, underwater, or overhead. Similar to cartridge systems, adhesive contained in the capsule is made up of multiple parts that must be mixed together. Capsule systems are unlike cartridge systems where the adhesive is mixed as it is being dispensed from the cartridge and placed in the hole using a nozzle. An adhesive capsule is a portion-controlled anchor system that is placed in the hole and then mixed either by impact, in the case of drive-in systems (Fig. 7.1a), or a combination of spinning and hammering of the anchor induced by a rotary hammer drill (Fig. 7.1b), in the case of spin-in systems. The glass or foil package that contains the adhesive material is mixed with the adhesive and remains dispersed within the mixed adhesive in the hole. The presence of the glass or foil in the adhesive does not impact the performance of the adhesive because its presence is part of the design of the product. The volume of the capsule matches exactly the corresponding hole. There is no flexibility with regard to hole diameter or hole depth.

The use of capsule adhesive anchor systems is not assigned to any specific application other than for bonding either threaded rod anchors or dowels to concrete substrates. This type of anchor is used at embedment depths that correspond to either a single capsule length or two-capsule lengths measured end-to-end. The type of packaging/delivery system and steel selection lends itself to applications where only a single or few anchors need to be installed.

Materials

Adhesive: Capsule adhesive material can be packaged in glass or foil containers (Fig. 7.2). Flexible foil and glass capsules need to be handled and stored with care. Capsules that have expired, or are cracked, cloudy, or discolored should not be used. A liquid resin (acrylic or vinylester), various accelerators, mineral aggregate, or cement are contained within the capsule (Fig. 7.3).[7] Similar to cartridge adhesives, capsules should be stored in a cool, dry, dark, well-ventilated location and not be exposed to sunlight. Refer to the MPII (Fig. 7.4) or

Fig. 7.1a: Inserting a glass capsule drive-in anchoring system after hole cleaning (Courtesy of Simpson Strong-Tie)

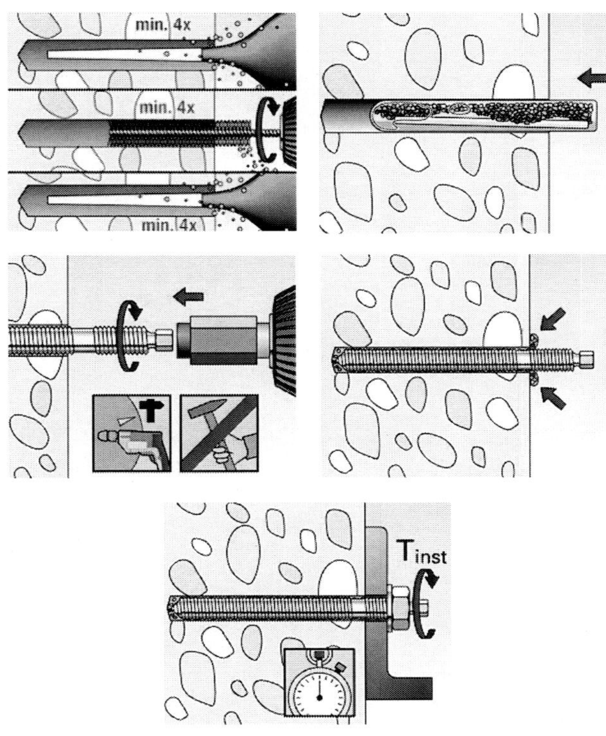

Fig. 7.1b: Installation of a glass capsule spin-in anchoring system (Courtesy of fischerwerke)

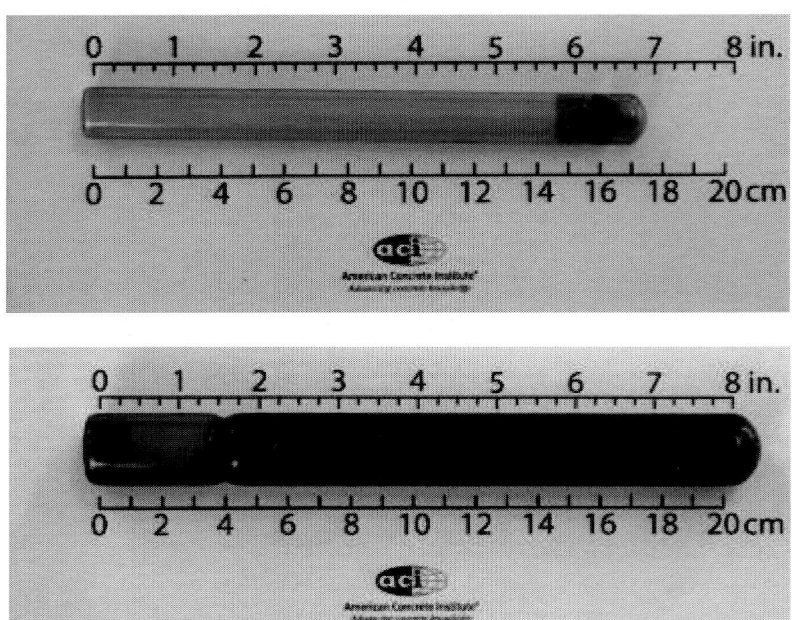

(a) Glass capsule

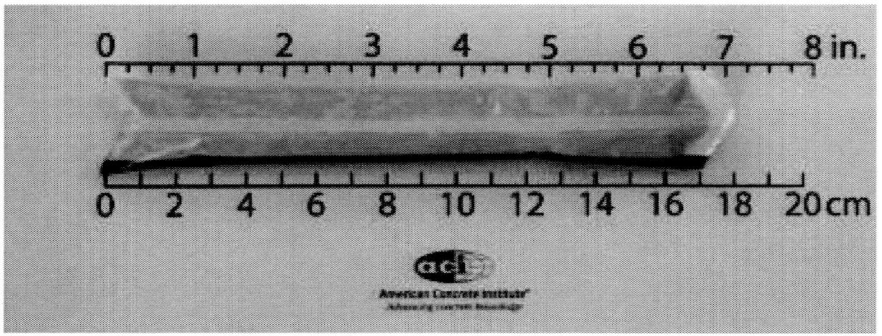

(b)   Foil   capsule

Fig. 7.2: Adhesive capsule systems contain an unmixed adhesive or other compounds (resin and hardener) and mineral fillers. These capsules have an exterior that is either: (a) glass; or (b) foil

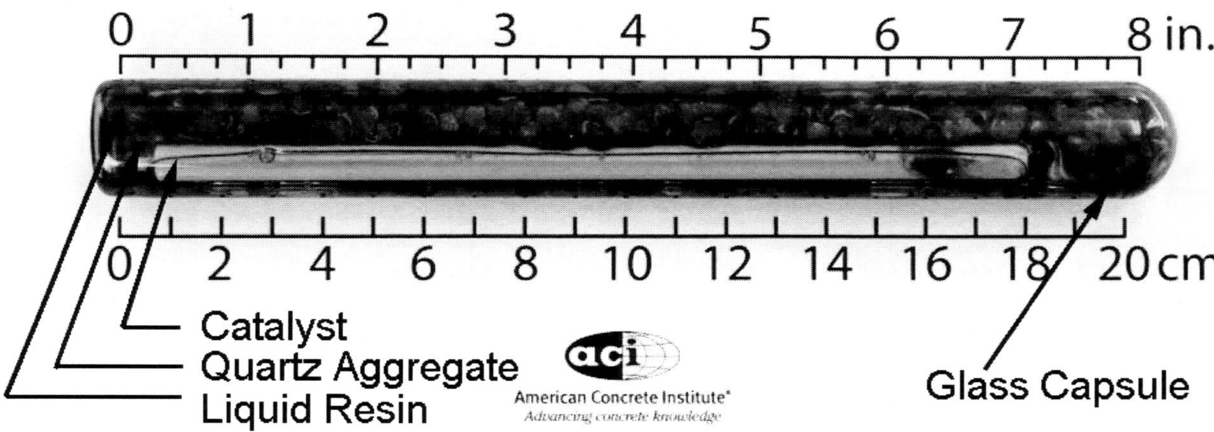

Fig. 7.3: Photograph of a glass capsule adhesive

Adhesive Anchoring Installation Instructions

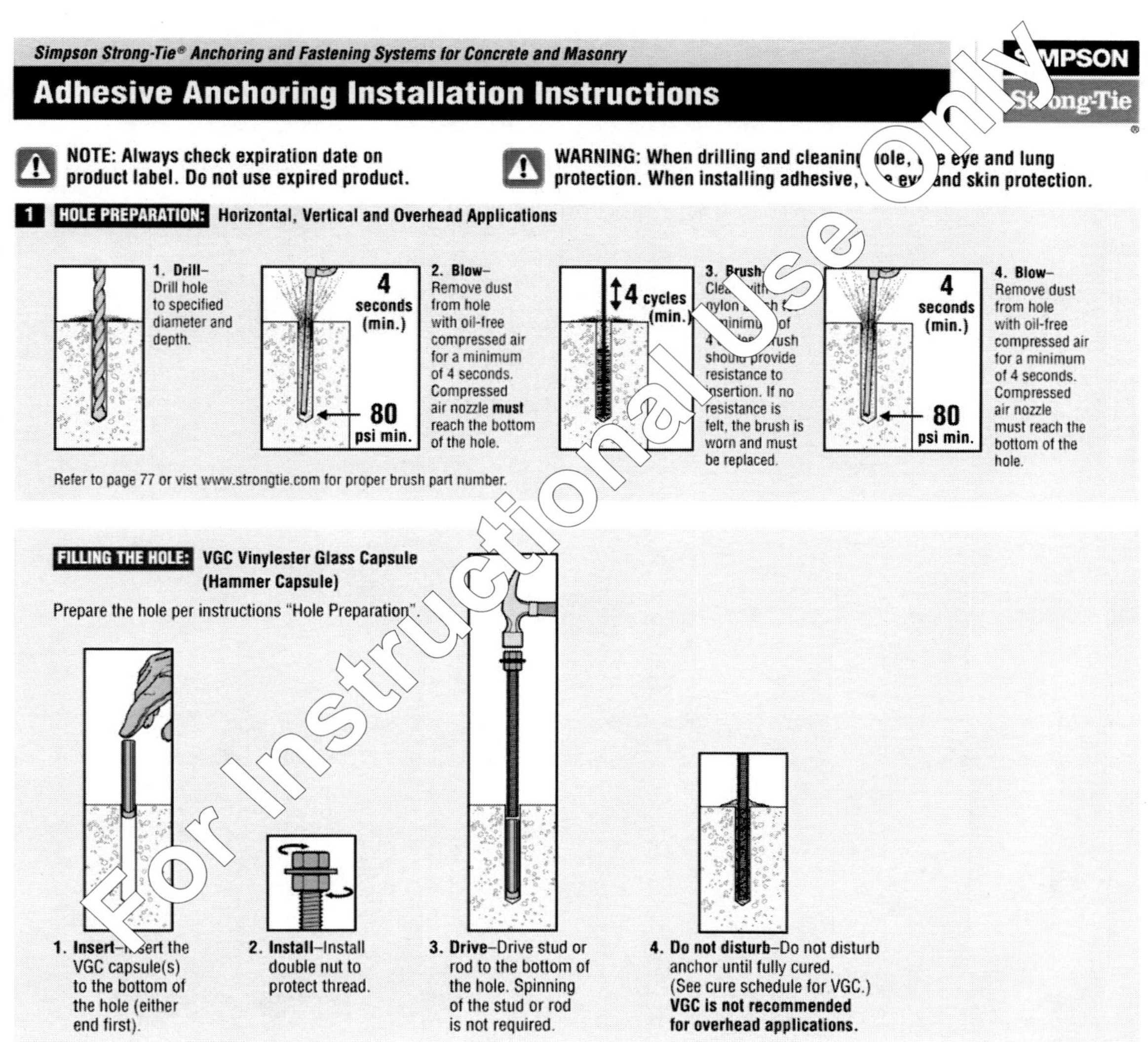

Fig. 7.4: Examples of MPII from various capsule adhesive product manufacturers. Do NOT use these for actual field installation. Refer to the MPII from the product that will actually be installed. (Courtesy of Simpson Strong-Tie, Powers Fasteners, and Hilti)

product packaging for proper storage conditions.

Capsule systems can be difficult to use in overhead installations because the adhesive when mixed is very flowable and can run out of the overhead hole unless retained. Additional equipment and instruction are needed for applications that are upwardly inclined. Many MPII do not recommend using capsule adhesive systems in an upwardly inclined orientation.

The MPII should indicate the proper capsule size for a specific hole diameter and depth. For deep embedment depths, it may be necessary to use multiple capsules in a single hole, only if this practice is permitted by the MPII.

Concrete with internal voids presents an installation problem if the voids are large enough to require a significant amount of adhesive. An example is manufactured voids in a precast structure or excessive embedment from drilling too deep. An indication of potential internal voids is a lack of excess adhesive at the surface of the concrete after anchor insertion. If the adhesive does not fill the hole after the anchor is inserted, the anchor cannot achieve the intended performance and the supervisor must be informed immediately.

Anchors: The anchor rods used with capsule systems often have a chiseled end (Fig. 7.5 and Fig. 1.5b) to puncture the capsule. Some products, however, may be designed for use with straight cut anchors. The relevant information is given in the MPII.

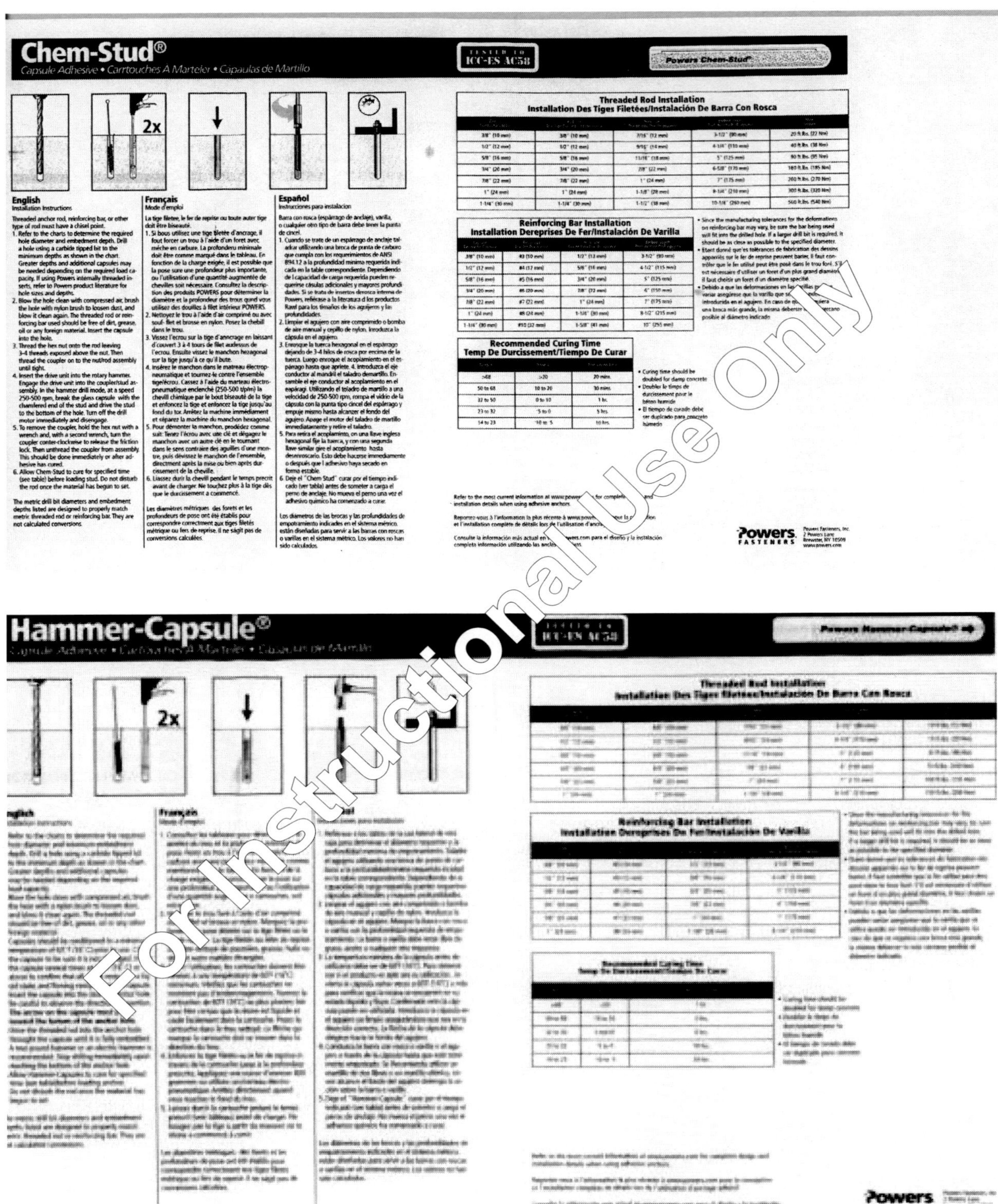

Fig. 7.4 - continued

Approvals (HVU)

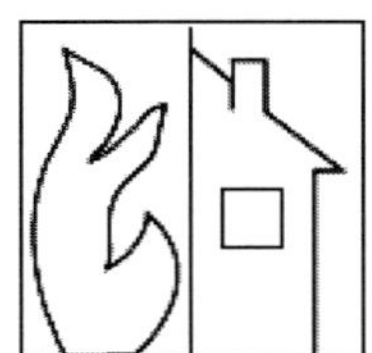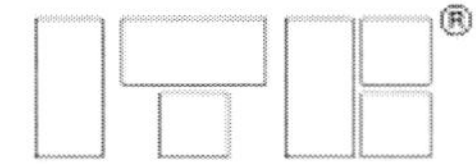

– Zulassungen sind vorrangig zu beachten.
– Approvals take precedence.
– Les agréments sont applicables en priorité.
– Normas tienen prioridad.

2

Fig. 7.4 - continued

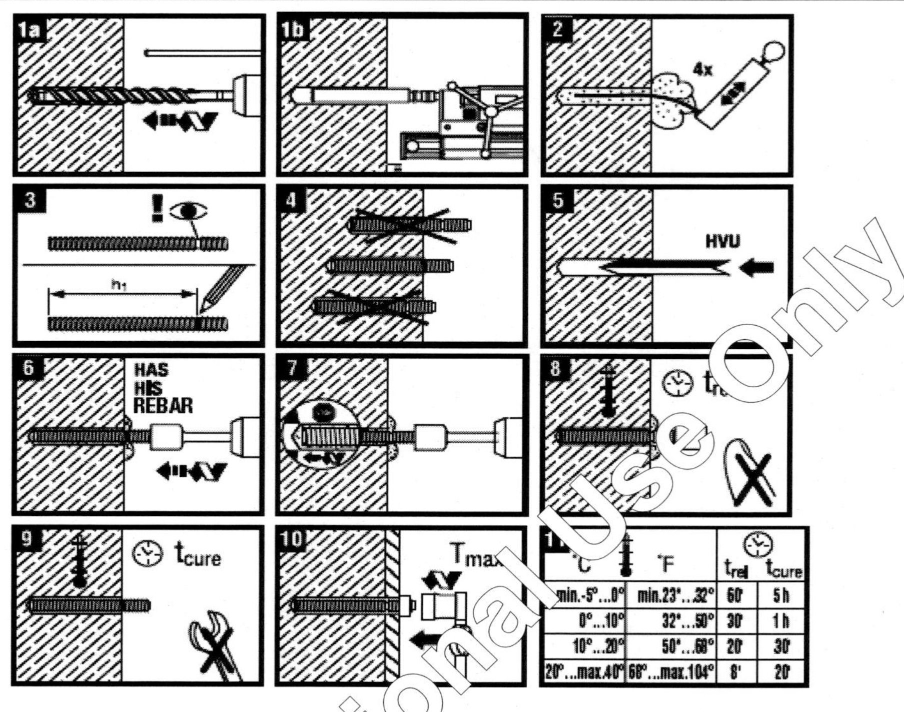

HVU + HAS(-R) HAS-HCR   HVU + HAS-E(-R)/(-F)

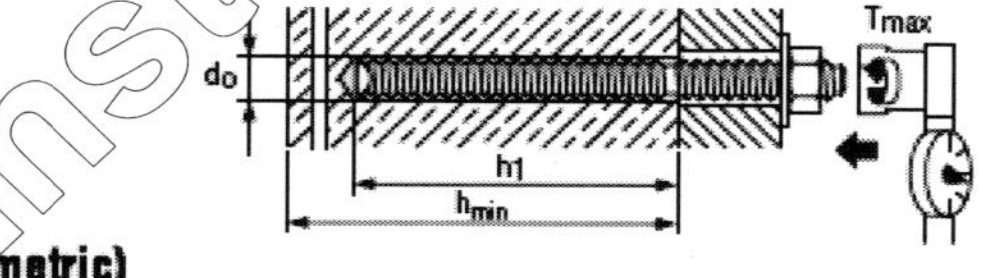

Europe + Asia (metric)

Setting Details / Anchor Size		M8	M10	M12	M16	M20	M24	M27	M30	M33	M36	M39
d_0	mm	10	12	14	18	24	28	30	35	37	40	42
h_1 (exact)	mm	80	90	110	125	170	210	240	270	300	330	360
h_{min}	mm	110	120	140	170	220	270	300	340	375	410	450
T_{max}	Nm	10	20	40	80	150	200	270	300	330	360	390
	TE-	1...30	1...30	1...30	1...60	50...60	50...80	60...80	60...80	60...80	60...80	60...80
	DD-							80...250	80...250	80...250	80...250	80...250

USA (inch)

Setting Details / HAS Rod Size			in.	3/8	1/2	5/8	3/4	7/8	1	1 1/4
			(mm)	(9.5)	(12.7)	(15.9)	(19.1)	(22.2)	(25.4)	(31.8)
d_0			in.	7/16	9/16	11/16	7/8	1	1 1/8	1 3/8
h_1			in.	3 1/2	4 1/4	5	6 5/8	6 5/8	8 1/4	12
			(mm)	(90)	(110)	(125)	(170)	(170)	(210)	(305)
h_{min}			in.	5 1/2	6 1/4	7 1/8	8 1/2	8 1/2	10 1/2	15
			(mm)	(140)	(159)	(181)	(216)	(216)	(267)	(381)
T_{max}	HAS-(E) HAS-(E)-Super HAS-(E)-SS		ft lb	18	30	75	150	175	235	400
			(Nm)	(24)	(41)	(102)	(203)	(237)	(318)	(542)

Fig. 7.4 - continued

Setting methods HAS

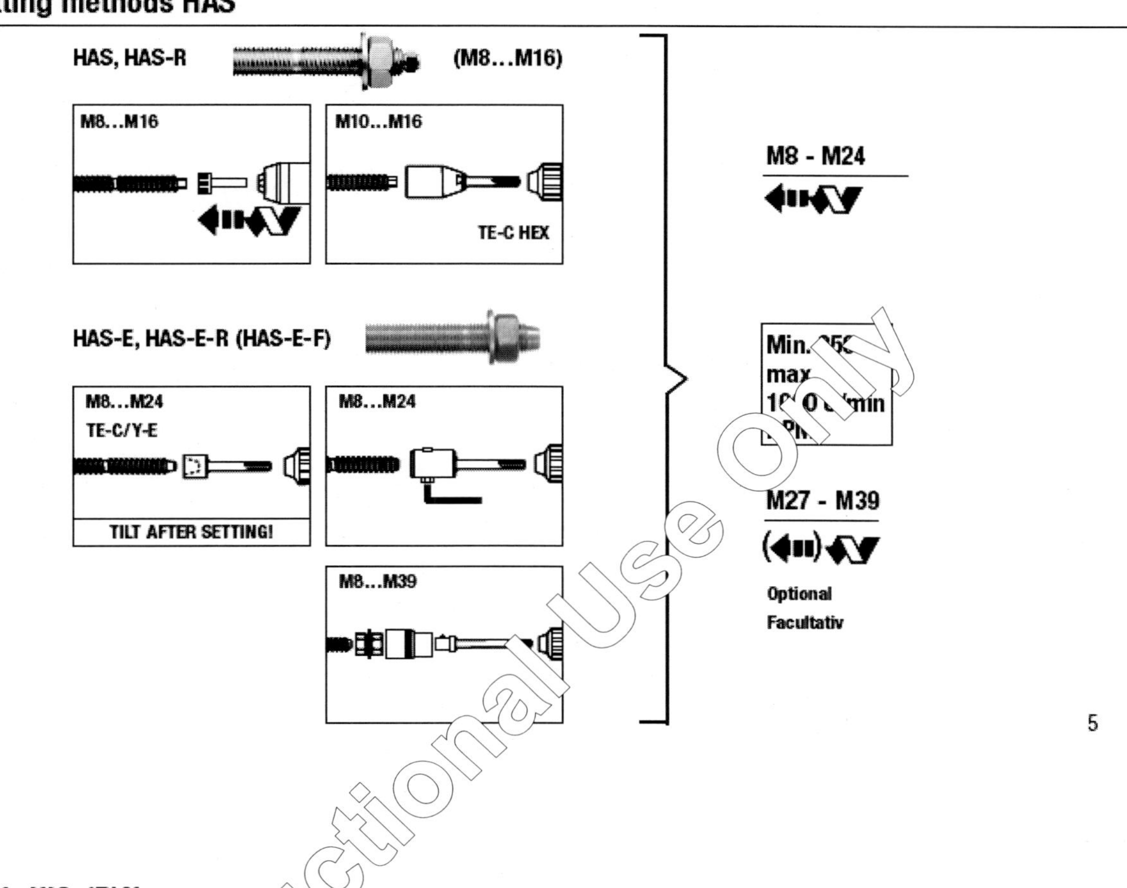

HVU+HIS-(R)N

HIS-(R)N			M8	M10	M12	M16	M20
HVU			M10x90	M12x110	M16x125	M20x170	M24x210
d_0		(mm)	14	18	22	28	32
h_1 (exact)		(mm)	90	110	125	170	205
h_{min}		(mm)	120	150	170	230	270
h_s min.-max.		(mm)	8-20	10-25	12-30	16-40	20-50
T_{max}	ETA	(Nm)	10	20	40	80	150
	TE-		1…40	1…40	1…40	40…80	40…80
			min. 250 - max. 1000 U/min RPM				

HIS Insert	in.	3/8	1/2	5/8	3/4
Details	(mm)	(9.5)	(12.7)	(15.9)	(19.1)
HVU		$1/_2$ x $4^1/_4$	$5/_8$ x 5	$7/_8$ x $6^5/_8$	1 x $8^1/_4$
d_0	in.	$^{11}/_{16}$	$7/_8$	$1^1/_8$	$1^1/_4$
h_1 (exact)	in.	$4^1/_4$	5	$6^5/_8$	$8^1/_4$
	(mm)	(110)	(125)	(170)	(210)
h_s max.	in.	1	$1^3/_8$	$1^1/_2$	2
	(mm)	(25)	(30)	(40)	(50)
T_{max}	ft lb	18	30	75	150
	(Nm)	(24)	(41)	(102)	(203)
h_{min}	in.	$6^3/_8$	$7^1/_2$	10	$12^3/_8$
	(mm)	(162)	(191)	(254)	(314)
	TE-	1…40	1…40	40…60	40…80
		min. 250 - max. 1000 U/min RPM			

6

Fig. 7.4 - continued

HVU+HIS-(R)N

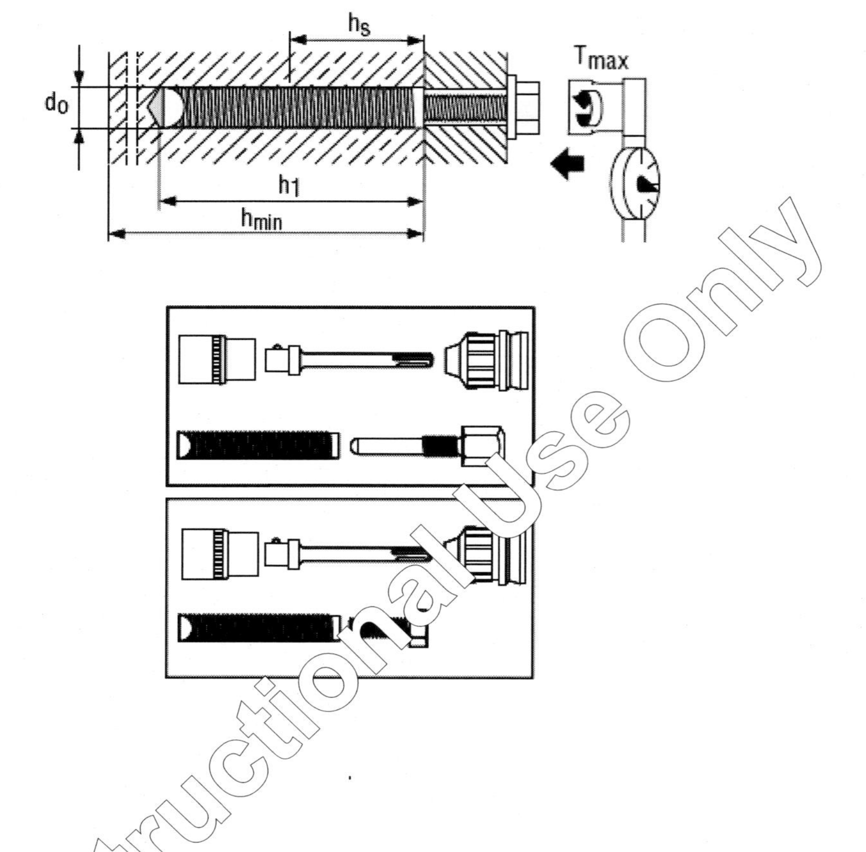

HVU+REBAR

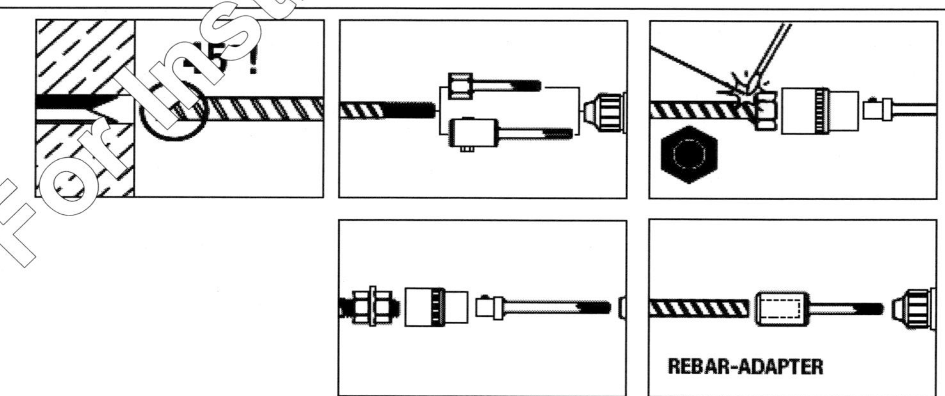

REBAR

(mm) HVU…	ø 10 10x90	ø 12 12x110	ø 14 16x115	ø 16 16x125	ø 20 20x170	ø 25 24x210	ø 28 30x270	ø 32 33x330	ø 36 39x360
d_0 (mm)	12	15	18	20	25	30	35	40	42
h_1 (mm) (exact)	90	110	125	125	170	210	270	300	360
h_{min} (mm)	120	140	170	170	220	270	340	380	460
TE-	1…30	1…30	1…60	1…60	50…80	60…80	60…80	60…80	60…80
DD-							80…250	80…250	80…250

Fig. 7.4 - continued

Hilti HVU adhesive anchor capsule

Read the instructions for use and safety instructions before using this product. Suitable for anchor fastenings in concrete with a minimum compressive strength of 15 N/mm².
Approvals requirements must be observed! Use only undamaged capsules!
Expiry date: See date printed on box and capsule. Do not use if expiry date has been exceeded!
Base material temperature: Must be between –5 °C/23 °F and 40 °C/105 °F when setting the anchor.
For any application not covered by this document, contact Hilti.

1 **Drilling the hole:** a.) Rotary hammer drill: Set the depth gauge to the correct drilling depth.
  b.) Diamond coring. Mark the correct drilling depth on the height adjustment mechanism.

2 Clean the hole immediately before setting the anchor. Remove drilling dust and standing water **from the base of the hole** by blowing out well with at least 4 strokes of the blow-out pump, or using compressed air or an industrial vacuum cleaner. The anchor holes must be free of dust, water, ice, oil, bitumen, chemicals or any other foreign matter or contaminants. **Poorly-cleaned holes = poor hold.**

3 Ensure that the specified setting depth is marked on the anchor rod. If not, add an embedment mark for example with tape or marker.

4 **Caution!** Check that the hole is drilled to the correct depth before setting the anchor. Hole depth is correct when the anchor rod contacts the base of the hole and the setting depth mark coincides with the concrete surface.

5 Push the anchor capsule into the drilled hole.

6 Use the setting tool at a speed of 250–1000 r.p.m. to drive the anchor rod into the hole, applying moderate pressure and with the hammering action switched on.

7 Switch off the rotary hammer drill immediately when the setting depth is reached (refer to mark on the anchor rod). **After setting, adhesive mortar must fill the annular gap completely, right up to the concrete surface. Caution!** Prolonged rotary action may cause mortar to be forced out of the hole, resulting in reduced anchor loading capacity.

8 The working time "**t rel**", which depends on base material temperature, must be observed (see fig. 11). The screwed-on setting tool may be removed only after the time "**t rel**" has elapsed.

9 After reaching the end of the working time "**t rel**", do not manipulate or disturb the anchor rod in any way until the curing time "**t cure**" has elapsed.

12

Safety regulations

10 A load may be applied to the anchor only after the curing time "**t cure**" has elapsed.
11 The working time "**t rel**" and curing time "**t cure**", which depend on base material temperature, must be observed!

Hilti accepts no liability for damage resulting from failure to observe the installation instructions, use of anchors of inadequate size, installation in base materials with inadequate load-bearing capacity or incorrect use of the product.

R 36	Irritating to eyes.
R 43	May cause sensitization by skin contact.
S 3	Keep in a cool place.
S 37/39	Wear suitable gloves and eye protection.
S 26	In case of contact with eyes, rinse immediately with plenty of water and seek medical advice.
S 28	After contact with skin, wash immediately with plenty of soap and water.

Xi irritant,
contains hydroxypropyl methacrylate,
contains dibenzoylperoxide

Expiry date: See date printed on the sales packaging (box).
Transport and storage instructions:
Store in a cool, dry and dark place only in the original packaging.
Transport temperature: –5 °C to 30 °C
Transport temperature (max. 2 days):   –20 °C to 40 °C
Storage temperature: 5 °C to 25 °C
If stored for a short period below –5 °C, warm up capsules prior to use for min. 1 day at 20 °C.

Disposal instructions:
Unusable foil capsules e.g. if expiry date has been exceeded, should be disposed of as hazardous waste material under observance of the official regulations.
EAK code no.: 20 01 28 or 08 04 10

13

Fig. 7.4 - continued

Hilti HVU adhesive anchor capsule

Read the instructions for use and safety instructions before using this product. Suitable for anchor fastenings in concrete with a minimum compressive strength of 15 N/mm².
Approvals requirements must be observed! Use only undamaged capsules!
Expiry date: See date printed on box and capsule. Do not use if expiry date has been exceeded!
Base material temperature: Must be between –5 °C/23 °F and 40 °C/105 °F when setting the anchor.
For any application not covered by this document, contact Hilti.

1 **Drilling the hole:** a.) Rotary hammer drill: Set the depth gauge to the correct drilling depth.
b.) Diamond coring. Mark the correct drilling depth on the height adjustment mechanism.

2 Clean the hole immediately before setting the anchor. Remove drilling dust and standing water **from the base of the hole** by blowing out well with at least 4 strokes of the blow-out pump, or using compressed air or an industrial vacuum cleaner. The anchor holes must be free of dust, water, ice, oil, bitumen, chemicals or any other foreign matter or contaminants. **Poorly-cleaned holes = poor hold.**

3 Ensure that the specified setting depth is marked on the anchor rod. If not, add an embedment mark, for example with tape or marker.

4 **Caution!** Check that the hole is drilled to the correct depth before setting the anchor. Hole depth is correct when the anchor rod contacts the base of the hole and the setting depth mark coincides with the concrete surface.

5 Push the anchor capsule into the drilled hole.

6 Use the setting tool at a speed of 250–1000 r.p.m. to drive the anchor rod into the hole, applying moderate pressure and with the hammering action switched on.

7 Switch off the rotary hammer drill immediately when the setting depth is reached (refer to mark on the anchor rod). **After setting, adhesive mortar must fill the annular gap completely, right up to the concrete surface. Caution!** Prolonged rotary action may cause mortar to be forced out of the hole, resulting in reduced anchor loading capacity.

8 The working time "**t rel**", which depends on base material temperature, must be observed (see fig. 11). The screwed-on setting tool may be removed only after the time "**t rel**" has elapsed.

9 After reaching the end of the working time "**t rel**", do not manipulate or disturb the anchor rod in any way until the curing time "**t cure**" has elapsed.

10 A load may be applied to the anchor only after the curing time "**t cure**" has elapsed.

11 The working time "**t rel**" and curing time "**t cure**", which depend on base material temperature, must be observed!

56

Safety regulations

Hilti accepts no liability for damage resulting from failure to observe the installation instructions, use of anchors of inadequate size, installation on base materials with inadequate load-bearing capacity or incorrect use of the product.

HVU Adhesive Capsules
For Industrial Use Only

Caution: Irritating to eyes and skin. May cause sensitization by skin contact. Contains silica sand.
Precautions: Avoid prolonged or repeated contact with skin and eyes. Wear suitable gloves and eye protection.
First Aid: For eye contact, flush with plenty of water while holding the eyelids apart. For **skin contact**, wash with soap and water.

See the Material Safety Data Sheet for this product	
Ingredient:	**CAS Number:**
Silicon dioxide (quarz sand)	14808-60-7
NJ Trade Secret Registry Nr.	19136100-5001p
NJ Trade Secret Registry Nr.	19136100-5004p
NJ Trade Secret Registry Nr.	19136100-5005p
Dicyclohexyl phthalate	00084-61-7
Dibenzoyl peroxide (< 2%)	00094-36-0

HMIRC registration number 4185,
granted 10 August, 2000

Transport and storage instructions:
Store in original packaging in a cool dry place and out of direct sunlight.
Transport temperature: - 5°C/23°F to 30°C/86°F
Transport temperature (max. 2 days):   -20°C/-4°F to 40°C/104°F
Storage temperature: 5°C/41°F to 25°C/77°F
If stored for a short period below -5°C/23°F, warm up capsules prior to use for min. 1 day at 20°C/62°F.

57

Fig. 7.4 - continued

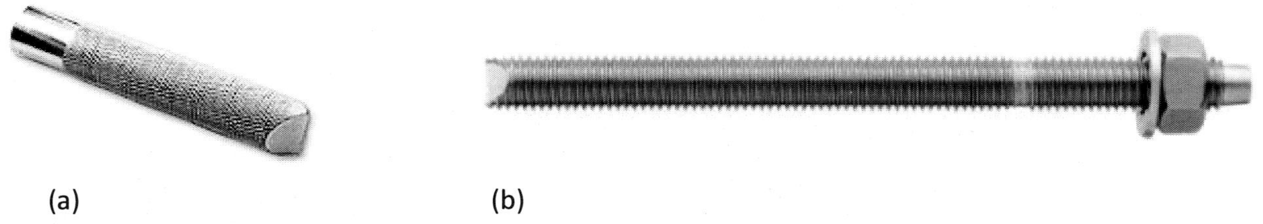

Fig. 7.5: Examples of anchors with chiseled ends: (a) internally threaded insert (Courtesy of Alma Bolt Company & Prime Fastener); and (b) a threaded anchor (Photograph courtesy of Hilti)

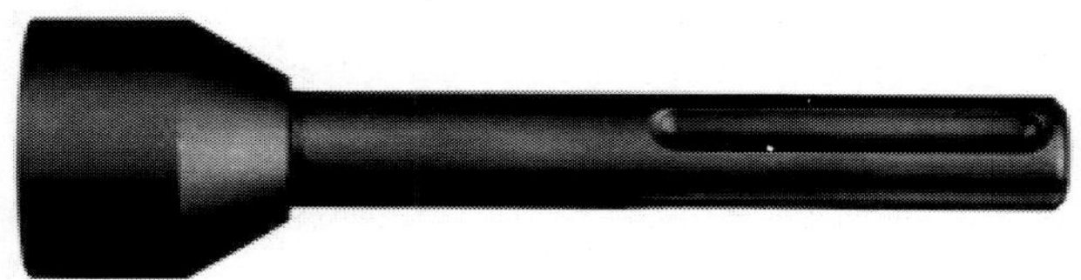

Fig. 7.6: Setting tool used to rotate/drill-in threaded anchor rod (Courtesy of Hilti)

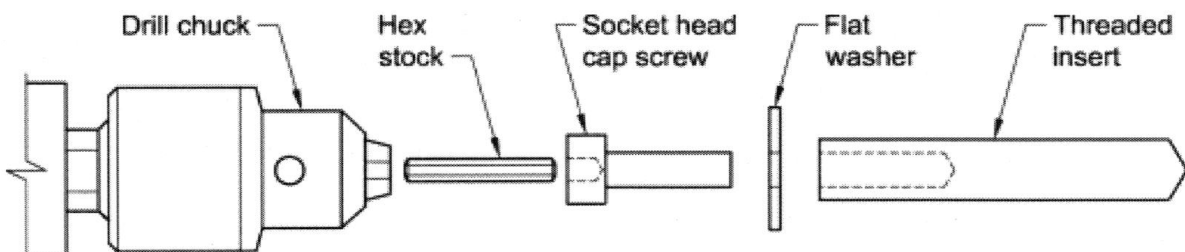

Fig. 7.7: Illustration of a socket head cap screw used as a driver for an internally threaded anchor

Because the anchor must be hammered or spun into the hole to mix the adhesive, care must be taken to not damage the threads that will remain exposed during the installation. The MPII may suggest the use of a washer and one or more nuts or a setting tool (Fig. 7.6) to protect the threads.

Equipment

After drilling the embedment hole with a rotary hammer drill or other drill types in the case of spin-in anchors, the drill may also be needed for setting the anchor. Depending on the specific product, the anchor may need a setting tool adaptor to install the anchor in either a rotary or hammer mode. The end of the setting tool will need to be compatible with the chuck on the drill. The corresponding information is given in the MPII.

Setting tools (Fig. 7.6) vary depending on the anchor system (threaded rod or reinforcing bar) and the manufacturer. Always check the MPII for details. Typical accessories will include drivers for threaded rod couplers and internally threaded insert tools that are used with the drivers for threaded rod. For internally threaded inserts, a socket head cap screw, washer, and short length of hex stock can be used as a driver (Fig. 7.7).

At the exposed end of a threaded rod anchor, it can be helpful to use a washer tightened between two nuts to "lock" the washer in place (Fig. 7.8). For threaded rod anchors that are driven into the hole, this arrangement will protect the threads on the end of the anchor from being damaged by the impact device. To protect the threads, the exposed end of the rod must be located inside the nut a small distance. When driving the threaded anchor, this arrangement allows the hammer to strike the nut instead of the threads. For anchors that are spun in, this arrangement also allows the use of a socket to spin the rod.

Procedure: Capsule Installation

The safety guidelines provided by the adhesive anchor system manufacturer should be read by the installer and strictly followed. PPE typically recommended or required for this operation includes safety glasses and gloves. Be sure your work area is properly ventilated.

To install a capsule adhesive anchor, the drilling, hole preparation, and cleaning procedures

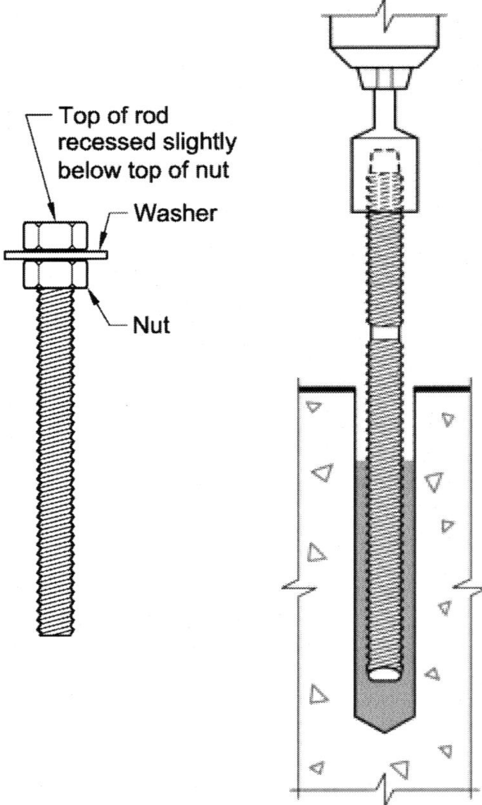

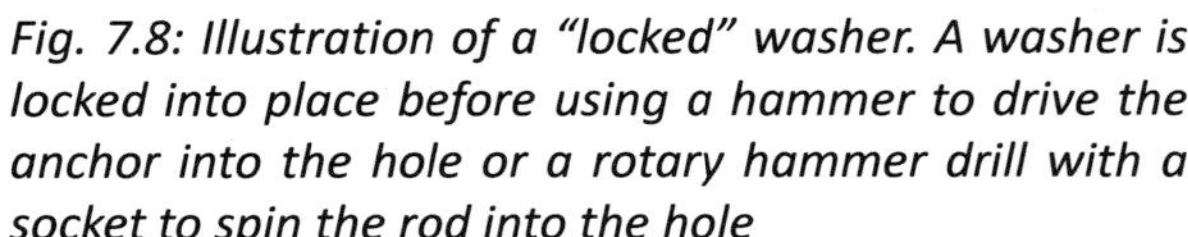

Fig. 7.8: Illustration of a "locked" washer. A washer is locked into place before using a hammer to drive the anchor into the hole or a rotary hammer drill with a socket to spin the rod into the hole

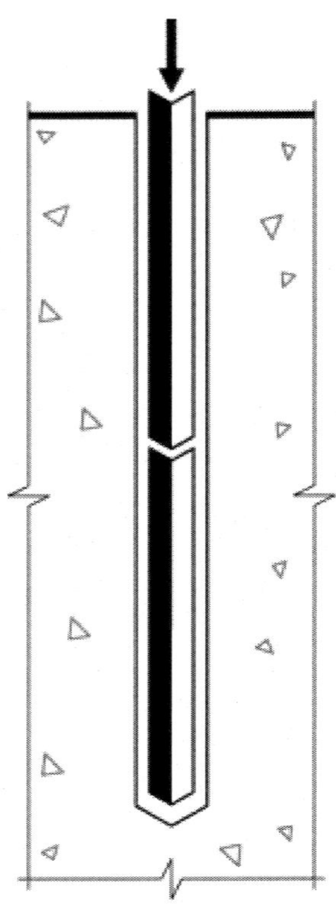

Fig. 7.9: A multiple capsule installation. Note that both capsules are oriented with the arrow pointing toward the deepest part of the hole. It is important to position the capsule as directed in the MPII

(blow, brush, blow) are similar to those where the adhesive is in a cartridge and mixed in the mixing nozzle. Be sure to consult the MPII for details on the hole cleaning requirements. Some capsule systems may not require extensive brushing to remove drilling dust from the sides of the hole because the dust will be scraped away by the mixing action as the anchor is installed.

Observe the storage conditions required by the MPII for the capsule, check the expiration date on the box containing the capsules, examine the exterior of the capsule for any punctures or visible discolorations, and make sure the appropriate installation equipment is available. The capsules must be stored at temperatures within the range stated in the MPII and not be exposed to sunlight.

Immediately prior to placing the capsule into the drilled and cleaned hole, the capsule should be checked to confirm that the resin has not hardened. Rigid, glass capsules require inversion several times to confirm that the adhesive materials are in a liquid state. Foil capsules can be checked by gently squeezing the capsule to ensure the contents are pliable and the adhesive components have not reacted and hardened.

Check the condition of the anchor and mark the embedment depth per the MPII for an anchor installed using a cartridge adhesive system. Prior to inserting the capsule, place the anchor in the hole to confirm the proper depth.

A single capsule (unless the hole depth requires more than one capsule) is placed in the hole. Depending on the product, there may be a correct direction to insert the capsule in the hole; however, some capsules are bi-directional, meaning that either end can be inserted into the hole first. Check the MPII for any direction requirements. In most glass capsules, an air bubble is entrapped. The MPII may require the capsule to be inserted into the hole with the bubble at the end closest to the concrete surface. In general, even if it is not required by the MPII, it is recommended to insert the capsule with the bubble closest to the concrete surface. The product manufacturer provides information on the size and number of capsules necessary to fill different size holes (Table 7.1). Review the MPII for the required number of capsules, and do not use portions of capsules. If the capsules are not bi-directional, be sure to position multiple capsules with the bottom end of each capsule pointed toward the deepest part of the hole (Fig. 7.9).

There are two methods for inserting the anchor and mixing the adhesive. The MPII will direct you as to which to use.

1. For an anchor that is driven into the hole, use a hammer or rotary hammer drill (in hammering mode only) to drive the anchor into the hole, stopping when the anchor touches the bottom of the hole.

2. For an anchor that requires spinning into the hole, position the capsule in the hole and use a rotary hammer drill and an anchor with a chiseled point to mix the components and install the anchor. Check the MPII for details on the connection of the anchor to the drill. Place the end of the anchor on the top of the capsule and drill down through the capsule. The MPII may specify a drill speed in revolutions per minute (RPM). Be careful to not over mix the material with the drill. Excessive mixing can lead to product "bleeding" out of the hole which will reduce the bond strength and performance of the anchor. You must perform mixing of the capsule systems in strict accordance with the MPII. Once the anchor has reached the bottom of the hole, the drilling action should be immediately stopped and the anchor should be detached from the drill without disturbing the position of the anchor in the hole. The hex nut is then held with a wrench to unthread the coupler. When using reinforcing bar, release the set lever and slide the coupler off the bar.[8] Do not over spin the anchor during the installation. This will cause adhesive to exit the hole, severely impacting anchor performance.

Differences between cartridge and capsule systems

Similar to cartridge adhesive installations, it is advisable to repeat the MPII cleaning procedures prior to installation if a hole has been "open" for a period of time and not protected.

Bond development along the embedment depth is limited by the quantity of adhesive in a capsule that fills the hole. Do not use part of a capsule. Attempting to do so will result in improper proportions of the materials. Therefore, the engineer should not specify an adhesive anchor capsule installation that does not have a product suitable for the required embedment depth. The advantage of the cartridge system is that

Table 7.1: Table from manufacturer's product literature. For a 3/8 in. diameter rod, depending on the required embedment depth, the number of capsules installed can be one or two; no portions of capsules (Courtesy of Hilti)

Rod Diameter in (mm)	Embed. Depth in (mm)	Adhesive Capsule(s) Required
3/8 (9.5)	3-1/2 (90)	(1) 3/8 x 3-1/2
	5-1/4 (133)	(2) 3/8 x 3-1/2
	7 (178)	(2) 3/8 x 3-1/2
1/2 (12.7)	4-1/4 (110)	(1) 1/2 x 4-1/4
	6-3/8 (162)	(1) 1/2 x 4-1/4 & (1) 3/8 x 3-1/2
	8-1/2 (216)	(2) 1/2 x 4-1/4
5/8 (15.9)	5 (125)	(1) 5/8 x 5
	7-1/2 (184)	(1) 5/8 x 5 & (1) 1/2 x 4-1/4
	10 (254)	(2) 5/8 x 5
3/4 (19.1)	6-5/8 (170)	(1) 3/4 x 6-5/8
	10 (254)	(1) 3/4 x 6-5/8 & (1) 1/2 x 4-1/4
	13-1/4 (337)	(2) 3/4 x 6-5/8
7/8 (22.2)	6-5/8 (170)	(1) 7/8 x 6-5/8
	10 (254)	(2) 3/4 x 6-5/8
	13-1/4 (337)	(2) 7/8 x 6-5/8
1 (25.4)	8 1/4 (210)	(1) 1 x 8-1/4
	12-3/8 (314)	(2) 7/8 x 6-5/8
	16-1/2 (419)	(2) 1 x 8-1/4
1-1/4 (31.8)	12 (305)	(1) 1-1/4 x 12
	15 (381)	(1) 1-1/4 x 12 & (1) 1 x 8-1/4
	18 (457)	(1) 1-1/4 x 12 & (2) 1 x 8-1/4

the quantity of adhesive can be adjusted by the installer to adequately fill the hole two-thirds full, whereas for a capsule system, you can use one or more complete capsules, but you cannot use half of a capsule.

Unlike cartridge adhesive installations, gel time is not as critical of an issue for capsule systems. This is because the anchor mixes the adhesive while being inserted so there is no delay between placement of the adhesive and insertion of the anchor. With this system it is important to not disturb the anchor during release of the setting tool.

Techniques for installing the anchor in a capsule anchor system require a hammer or rotary hammer drill. However, in the cartridge system, the anchor is simply placed into the hole by hand and slightly twisted per the MPII.

Cited References

[8] Powers Fasteners, "Product Information for Chem-Stud$^\lozenge$ Spin-Type Capsule Adhesive Anchor System," www.powers.com/pdfs/chemical/6504.pdf

---

Checklist for Capsule Anchors:

- Use PPE
- Review MPII
- Drill and clean the hole using procedures similar to those for cartridge adhesive systems
- Check hole depth
- Verify embedment depth
- Verify perpendicularity
- Inspect capsule
- Insert capsule
- Insert anchor by driving or spinning it into the hole per MPII
- Protect and support the anchor until the adhesive has fully cured
- Verify the length of exposed anchor

Chapter 7

1. __________ adhesive anchor systems are an alternative to cartridge anchor systems. Both contain adhesives made up of multiple compounds.

   A. Capsule
   B. Dispenser
   C. Nozzle
   D. Retaining cap

2. Capsule adhesives are mixed by __________.

   A. a paddle mixer
   B. a static mixing nozzle
   C. impact or spinning of the anchor
   D. stirring and then pouring

3. Consult the ______ for proper storage conditions for capsule adhesive anchors.

   A. contract drawings
   B. MPII
   C. MSDS
   D. PPE

4. Some capsules are ________________, meaning that either end can be inserted into the hole. Others are not and should be positioned correctly.

   A. bi-directional
   B. glass
   C. foil
   D. rigid

5. For long embedment depths, it may be necessary to use multiple whole capsules in a single hole, for details refer to the MPII.

   A. True
   B. False

6. Immediately before use, check a foil capsule to ensure that it is pliable and the adhesive has not hardened by ________________ it.

   A. gently squeezing
   B. slowly inverting
   C. rapidly opening
   D. vigorously shaking

7. Given the manufacturer information provided below, for a capsule adhesive anchor installation with anchor diameter of 5/8 in., and an embedment depth of 10 in. How many adhesive capsule(s) are required?

Rod Diameter, in. (mm)	Embed. Depth, in. (mm)	Adhesive Capsule(s) Required
5/8 (15.9)	5 (125)	(1) 5/8 x 5
	7-1/2 (184)	(1) 5/8 x 5 & (1) 1/2 x 4-1/4
	10 (254)	(2) 5/8 x 5

A. (1) 5/8 x 5
B. (1) ½ x 4-1/4
C. (1) 5/8 x 5 & (1) ½ x 4-1/4
D. (2) 5/8 x 5

8. For capsule adhesive anchor installations, the anchor can be bent, wet, dirty and oily, with damaged threads.

A. True
B. False

9. Large voids in the concrete can be problematic for capsule adhesive installations as they may require a significant amount of additional adhesive.

A. True
B. False

10. Some capsule adhesive systems may not require ______________ as part of the hole cleaning process because the mixing action of the anchor will scrape away dust from the side of the hole as it is drilled or hammered in.

A. blowing
B. brushing
C. vacuuming
D. water-flushing

Chapter 8: Installation Problems

As indicated in previous chapters, successful adhesive anchor installation depends on creating the best concrete surface in the hole to which the adhesive can adhere and the cleanliness of the anchor. This requires the use of proper equipment, tools, and techniques. Materials and equipment from different systems must not be mixed because the materials and equipment are specifically designed by the manufacturer to work together as a system.

Some common and frequently encountered problems during the installation of adhesive anchors include: entrapped air pockets in the adhesive, an unclean drill hole, water in the drill hole, damp concrete, crooked or bent anchors, incomplete mixing of adhesive, holes too close to edges, dirty or oily anchors, varying weather conditions such as ice, snow, and temperature swings during installation, adhesive that is too cold or too hot, and the adhesive not bonding to the anchor. Typical contaminants encountered in the field are cutting oil residue and dirt; these prevent adhesive from bonding with the anchor or getting into or around the rod threads or bar lugs. We will discuss these problems and the preventative actions that the installer can perform during the installation process.

Drilling

Drill Hole Interference: Prior to drilling, have the current set of drawings available to locate the holes to be drilled and avoid drilling into reinforcement or other embedded items. Testing to locate embedded items may also have to be performed at the job site. If embedded items are accidently encountered during the drilling process, stop and notify your supervisor immediately. Metal filings, a change of color of the drilling dust, or an increase in drilling difficulty are indicators that the drill has possibly encountered post-tensioning tendons, reinforcing steel, conduits, or other embedded items.

Drill Hole Voids: If you observe a void below the concrete surface where an anchor is to be installed, stop and notify your supervisor immediately. Concrete with significant voids will require additional adhesive to fill the void before the anchor is installed (and may not be suitable for anchorage at all). As shown in Fig. 8.1, the thickness of the adhesive will also be greater in some areas. These factors may harm the performance of

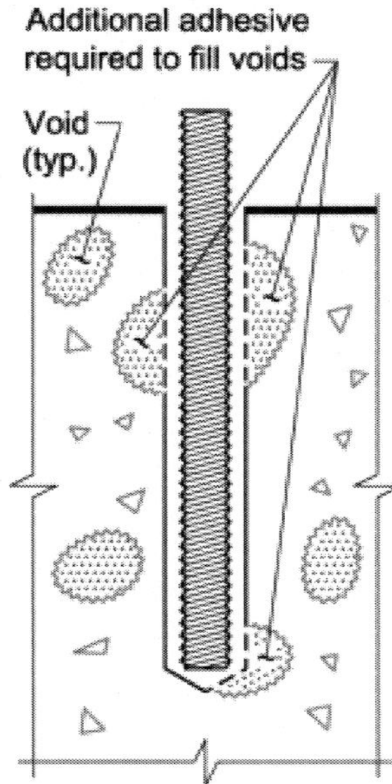

Fig. 8.1 Concrete with voids will produce a thicker adhesive layer than the layer used in qualification testing

the anchor. Similarly, when drilling into a slab-on-ground, you should stop and notify your supervisor immediately if the drill bit punches through the slab to the subgrade below.

Drill Hole Depth Variations: If one anchor in a multiple-anchor installation extends above the concrete more than the others, the hole may have been drilled too shallow or the anchor may not have been completely inserted into the hole to the proper embedment depth. If the hole was drilled too shallow, the anchor will need to be removed. If the adhesive has already been injected into the hole before it is discovered that the hole is too shallow, the anchor should be removed before the working time of the adhesive passes. After the adhesive cures, the hole will need to be redrilled. For this reason, it's very important to check the depth of the hole before injecting the adhesive. If the hole was drilled short because an embedded item has prevented drilling to the required depth, notify your supervisor. Using anchors of equal length will aid in the effectiveness of this visual check. The project engineer will need to be consulted to determine if and where the anchor may be relocated, or an alternative solution such as steel extender plate could be used to distribute load over a larger area.

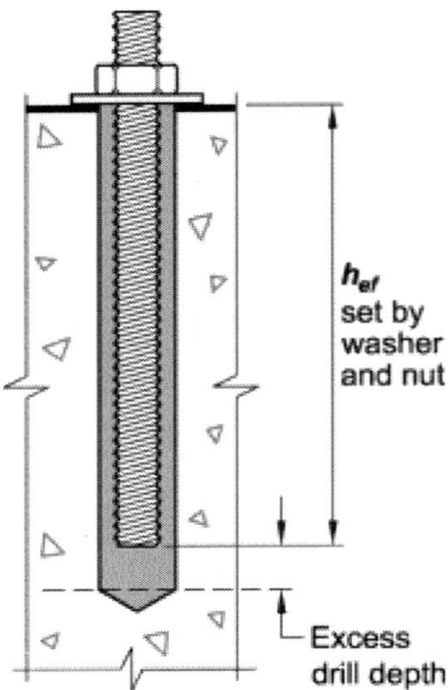

Fig. 8.2 Using a washer and nut to set the embedment depth in a hole that was drilled too deep

Fig. 8.3: Anchor alignment exceeds the three degree limit

If the hole was drilled too deep but did not completely drill through the concrete, a washer and nut set at the proper depth can be used to keep the anchor positioned at the correct embedment depth (Fig. 8.2). A larger quantity of adhesive may need to be injected into the hole to fill the extra volume of the hole. However, it must be noted that this remedy will not work for capsule systems. Caution should be used to avoid over-drilling the holes past the specified embedment depth, as the concrete thickness may not be known and an increased hole depth may weaken the connection.

Worn Drill Bit: One consequence of using a poor/worn drill bit can be an elliptically (oval) shaped hole. Because the anchors or reinforcing bar are round and sometimes cut on a bar shear that significantly deforms the sheared end, it may be difficult to fit a round anchor into an elliptical hole. To prevent drilling elliptical holes, check the wear mark on the drill bit frequently and replace the bit when appropriate. An anchor can be inserted into the hole before adhesive is injected to verify fit of the anchor. It is advisable to do this step regardless of the shape of the hole. Another consequence is that the hole will be undersized. That is, the diameter of the hole won't be large enough for the adhesive and anchor diameter being used. This too, may unfavorably impact the installation process and the anchor performance. Pay attention to wear sensors on the drill bits (if present) and perform visual inspections daily. Measurement of the bit diameter is recommended to confirm that the bit is within acceptable tolerances of ANSI

B212.15 (see Fig. 3.9 for an example).

Preparing the Hole

Unclean Holes: An unclean hole is a potential source for a complete failure of an anchor system. Anchor system adhesives achieve load carrying capacity by bonding to the concrete. If the adhesive does not have a clean, rough surface on which to bond, the possibility of bond failure increases. Cleaning of the hole should be performed in accordance with the MPII. Any dust, grease, cutting oil, ice, snow, or other contaminants that may line the surface of the hole are not acceptable. The cleanliness of the air supply used to blow out the holes is often overlooked. The source of air must produce clean air that is free of oil, antifreeze, and water. Using air sources such as gas-powered tools, modified leaf-blowers, or blowing into the hole through a straw are not acceptable options. They can introduce residue into the hole that could reduce the adhesive/concrete bond. Manufacturers published load values are based on installations using the proper equipment, air quality and pressures, and curing times as specified in the MPII.

Moisture in Holes: Water in the hole is not acceptable unless the product is qualified for this condition, which should be noted on the MPII and project specifications. Water in the hole or a damp hole can reduce the bond strength of some adhesive products. If you encounter this situation and you're not sure the product supplied can be used, ask your supervisor.

Anchor Alignment: Most installations require the anchor to be perpendicular to the concrete substrate. If the anchor is crooked (Fig. 8.3), it can affect the anchor's behavior in tension and shear loading. Additionally, it may make it difficult to install the fixture being anchored to the concrete. Generally, anchors should be installed within 3 degrees of perpendicular to the concrete surface, unless noted otherwise on the contract drawings.

Adhesive Mixing: Incomplete mixing of the adhesive that is injected into the hole will prevent the adhesive from achieving its full performance characteristics and will unfavorably impact the performance of the anchor system. To reduce the possibility of this potential installation problem, the adhesive anchor installer responsible for the installation must: follow the MPII provided, completely mix the adhesive, discard the initial adhesive prior to filling the hole, and ensure that the adhesive is properly cured. All cartridge-type products require wasting some of the initial adhesive that comes out of the nozzle. This step must never be skipped. For capsule systems, inspection of the adhesive capsule at the time of installation is required.

Temperature: Be aware of the temperature condition of the concrete, anchor, and the adhesive. Do not install adhesive systems at higher or lower temperatures than suggested in the MPII. High temperatures will make the adhesive set faster, reducing the working time and possibly leading to problems such as not being able to insert the anchor to the proper depth and correctly bond the anchor to the adhesive. Conversely, low temperatures can increase the working and cure time. This may pose a greater challenge in securing and protecting the installed anchor. Low temperatures may prevent some products from curing at all. If the installation temperature range is not listed on the MPII, it will be difficult to determine if and when the adhesive has cured to its full design strength. Use of infrared and contact thermometers to check hole and cartridge temperatures prior to installation is a good practice and provides feedback to the installer. Heating or cooling of the adhesive and anchoring material can be accomplished by placing it in a temperature controlled environment (such as a jobsite trailer). Heating or cooling of the concrete can be accomplished in a number of ways, but may not be practical. If this situation is encountered, review the installation procedures with your supervisor.

Cartridge adhesives

Never use a product in a damp, water-filled, submerged, or overhead application unless you can confirm the adhesive has been qualified for this use by the manufacturer. Follow procedures outlined in the MPII. Products developed and qualified for use in these unique applications typically have qualities that facilitate installation. For example, use a higher viscosity adhesive to reduce or prevent the adhesive from dripping or running out of horizontal and overhead holes.

Injecting the Adhesive into the Drill Hole: Entrapped air pockets are indicated by observing the following:

- the anchor element "springs back" out of the hole after insertion;
- the embedment depth mark on the installed anchor falls above the surface of the concrete; or
- audible snapping or popping noises as the anchor is inserted into the adhesive.

To avoid entrapping air voids in the adhesive, be certain to slowly withdraw the mixing nozzle and extension tube during filling from the deepest part of the hole to the surface. Always keep the nozzle or extension tubing just below the surface of the adhesive being injected. Maintain steady pressure on the adhesive dispenser trigger to keep adhesive consistently flowing from the nozzle. Erratic pumping can trap air between the adhesive and the sides of the hole. Keep the nozzle tip in contact with the surface of the hole and slightly buried in the adhesive. Insert the anchor into the hole with a slight rotational motion. Adhesives are generally very thick and have a surface tension characteristic that needs a chance to seep into the concrete and encapsulate the anchor. After installation and before the working time elapses, keep adjustments of the rod to a minimum. The start of working time begins when the mixed adhesive materials come out of the nozzle or extension tubing.

If it is discovered that an installation contains entrapped air, the only simple remedy is to remove the anchor, allow the adhesive to harden, then re-drill the hole and repeat the installation. This can be costly and delay the project; for this reason, special care needs to be taken when filling the hole with adhesive.

Inserting the anchor

Adhesive may not bond to the anchor if the anchor has contaminants on it. This is why an improperly installed anchor from one installation should not be reused in another installation. Additionally, improper placement of the anchor can reduce the bond strength of the installation. It's important to position the bottom of the anchor in contact with the deepest part of the hole, center it, and if required by the MPII, secure it until fully cured.

Always be aware of the final length of the exposed anchor. Protect the exposed anchor to avoid it becoming a safety hazard and so it does not get damaged. These steps will be very helpful to your supervisor and the steel erectors.

Capsule adhesives

For capsule systems, over-spinning the anchor will cause the adhesive to exit the hole and allow air entry into the hole. This is not desirable as the hole will not be full of adhesive, reducing the contact area at the two bond surfaces (concrete-adhesive and adhesive-anchor).

Proper adapters and spinning (revolution per minute), correct hole depth and hole size, and selection of an anchor with an appropriately shaped end should all be checked against the MPII for the product being installed. Stacking of capsules should only be performed when specifically permitted per MPII. As with cartridge systems, it's important to only use capsule systems in applications for which they have been designed.

Improper Installation

If you discover that an anchor was improperly installed after the adhesive has hardened, notify your supervisor. Resolutions of this problem may include proof loading the anchor or core drilling the anchor out and installing a new one. Your supervisor must decide the course of action. By following the MPII and the good practices discussed in this document, this situation can be avoided.

Chapter 8

1. If a hole was drilled too deep, but did not drill through the concrete member, the hole should __________.

    A. be used if a washer and nut set at the proper depth positions the anchor at the correct embedment depth
    B. be refilled with a repair grout and re-drilled to the proper depth
    C. be filled with adhesive and re-drilled to the proper depth
    D. not be used for an anchor installation

2. To prevent drilling misshapen holes, check the _________ on the drill bit frequently, and replace the drill bit when appropriate.

    A. flute
    B. shank
    C. tip
    D. wear mark

3. Wet or damp holes can _________ the bond strength of some adhesive products.

    A. increase
    B. reduce
    C. support
    D. toughen

4. A(n) ____________ hole is a potential source of failure for an anchor installation.

    A. narrow
    B. wide
    C. clean
    D. unclean

5. When the anchor "springs back" after being inserted or snapping noises are heard during insertion, it is likely that ___________.

    A. the hole is too shallow
    B. the hole is too deep
    C. there are no air pockets in the adhesive
    D. there are air pockets in the adhesive

6. In a capsule adhesive anchor installation, over-spinning the anchor will cause the adhesive to _________ the hole.

    A. exit
    B. fill
    C. level
    D. stay in

7. Anchors used in adhesive anchor installations should always be ___________________.

    A. clean
    B. dirty
    C. greasy
    D. wet

8. For cartridge installations, maintain __________ pressure on the adhesive trigger to keep a constant and consistent flow of adhesive from the nozzle.

    A. irregular
    B. light
    C. steady
    D. uneven

9. Complete mixing of an adhesive injected into a hole can prevent the adhesive from achieving its full performance characteristics.

    A. True
    B. False

10. Using an adhesive product designed for a vertical down installation in an overhead installation may not work. It is important to check the MPII and package labeling to be sure the product is suitable for the application.

    A. True
    B. False

PERFORMANCE EXAM

PART I – Installation of a Vertical-Down Adhesive Anchor

Preparation, Safety, Drilling, and Cleaning the Hole

1. Examinee reviewed the MPII and MSDS before beginning the installation?

2. Examinee selected the correct drilling equipment (e.g., drill setting, drill bit diameter, style and length, etc.) for the installation?

3. Examinee verified proper dust extraction equipment is available and in working order?

4. Examinee set depth gauge, marked drill bit, or used other methods to determine correct drilled-hole depth?

5. Examinee verified drill bit depth gage markings relative to dust extraction accessory used.

6. Examinee is wearing the appropriate personal protective equipment as detailed in the MPII.

7. Examinee set dust extraction accessory in proper location on concrete surface?

8. Examinee turned dust extractor unit on while drilling the hole?

9. Examinee took measures to ensure that the hole is drilled perpendicular to the concrete (tolerance: 3 degrees from perpendicular)?

10. Examinee drilled the hole to the specified depth of 4 inches (tolerance: +0.5 in./ 0 in.)?

11. Examinee selected equipment for hole cleaning in accordance with the MPII (e.g., brush diameter and type, compressed air equipment, manual blowout pump, etc.)?

12. Examinee used dust extraction accessory with a rubber lid in place during blowing out of hole procedure? (A used 4x8-in pad cap works best.)

13. Examinee turned dust extractor unit on during blowing out of the hole procedure?

14. Examinee cleaned the hole in accordance with the MPII (e.g. followed specified sequence and number of blowing and brushing steps, etc.)?

15. Why is it important to clean the hole?
    - ***To ensure the adhesive bonds to the concrete; or***
    - ***So the anchor performs as designed.***

16. If the anchor is to be installed the following day, what must be done?
    - ***Protect it to keep water and dirt out of the hole; or***
    - ***Protect it from contamination; or***
    - ***Re-clean the hole the next day.***

17. Examinee verified the proper hole depth prior to beginning injection of the adhesive?

Preparing and Injecting the Adhesive

18. Today is (Date), is this cartridge usable?
    - *Examinee responded appropriately, based on the adhesive's printed shelf life or expiration date?*

19. If this cartridge's temperature were 86 °F, would it be usable?
    - *Examinee responded "yes" appropriately based on the MPII?*

20. If this cartridge's temperature were 32 °F, would it be usable?
    - *Examinee responded "no" appropriately based on the MPII?*

21. Examinee properly assembled all of the components necessary for the adhesive injection (for the anchor size and depth to be installed)?

22. Examinee rejected the initial amount of adhesive dispensed from the mixing nozzle in accordance with the MPII?

23. Examinee demonstrated the correct use of the dispenser (start/stop/dispense rate of injection of adhesive)?

24. Examinee injected adhesive into the hole from the bottom of the hole in the manner specified by the MPII (approximately 2/3 full)?

Installing the Anchor

25. What is the anchor element checked for immediately before inserting it into the adhesive-filled hole?
    - *The anchor element is checked for "damage and contaminants"; or*
    - *It's "clean," "oil-free," etc.*

26. Examinee inserted the anchor in the manner specified by the MPII?

27. Were air pockets apparent during and/or following insertion of the anchor element?

28. Examinee cleaned excess adhesive from the concrete surface?

29. Was the adhesive left at the proper level / in the proper condition per the MPII?

Complete the following steps after the adhesive has reached the appropriate gel time.

30. Has final embedment depth been achieved (4 in. + 0.5 in./−0)?

31. Is perpendicularity within tolerance (3 degrees or less)?

Part II – Overhead Adhesive Injection

Injection into a sealed, clear 7/8-inch ID, rigid polycarbonate tube positioned overhead using a piston plug on the end of flexible extension tubing or similar device.

1. Today is (Date); is this cartridge usable?
   - ***Examinee responded appropriately based on the adhesive's printed shelf life or expiration date?***

2. Examinee properly measured the depth of the overhead hole and marked (on the extension tube with piston plug) the two-thirds depth?

3. Examinee properly used the provided extension tube with piston plug to inject the adhesive? (Piston plug was oriented correctly—in the hole first?)

4. Examinee rejected the initial amount of adhesive dispensed from the mixing nozzle in accordance with the MPII?

5. Examinee properly installed the extension tube with piston plug on the leading end, and then inserted the plug all the way to the top of the hole (tube)?

6. Examinee properly injected adhesive into the hole and allowed the piston plug to be displaced downward at a reasonably constant rate, by back pressure?

7. Examinee executed injection of adhesive into the overhead hole (tube) following all safety requirements of the MPII and MSDS?

8. Adhesive was injected to the proper depth of approximately two-thirds of the hole (tube) depth? (Examiner instructs Examinee to fill the remainder of the hole (tube) to eliminate the possibility of adhesive sagging into the air void in the bottom third. Examiner installs bottom cap in tube to hold adhesive and prevent sag.)

9. What is the next step?
   - ***The anchor rod or rebar is to be installed with a twisting motion and supported in place.***

10. Adhesive injection into hole (tube) was visibly evaluated to have a tolerable amount of air voids, or none at all?

Evaluation of cut tubes—Complete the following step after the adhesive has cured.

11. Longitudinal cross-sectioning of the polycarbonate tube revealed a tolerable amount of void(s) in the top (deepest) two-thirds of the adhesive injection pursuant to the grading rubric?

Chapter 1 – Introduction

Question	Correct Answer	Reference
1.	C	9
2.	A	9
3.	B	10
4.	C	11
5.	E	18
6.	A	18
7.	D	21
8.	C	9, 11, 21
9.	D	21
10.	A	12-14
11.	A	13
12.	A	10
13.	C	17-18
14.	A	18
15.	E	9
16.	D	13

Chapter 2 – Information for Proper Installation

Question	Correct Answer	Reference
1.	A	27
2.	F	27
3.	B	27
4.	D	37
5.	A	37
6.	C	38
7.	B	27
8.	B	27
9.	A	27
10.	D	36

Chapter 3 – Drilling the Hole

Question	Correct Answer	Reference
1.	A	41
2.	B	46
3.	D	46
4.	A	44
5.	B	44
6.	A	45
7.	A	44
8.	D	46
9.	B	43
10.	A	43

Chapter 4 – Preparing the Hole

Question	Correct Answer	Reference
1.	C	51
2.	A	51
3.	B	51
4.	B	51
5.	B	54
6.	A	53
7.	A	52
8.	D	53
9.	A	54-55
10.	A	54

Chapter 5 – Cartridge Systems – Injecting the Adhesive

Question	Correct Answer	Reference
1.	B	59
2.	D	59
3.	A	60, 62
4.	B	60-61
5.	B	61
6.	B	61-62
7.	C	61
8.	A	63
9.	A	63
10	C	64

Chapter 6 – Inserting the Anchor (Cartridge Systems)

Question	Correct Answer	Reference
1.	A	69
2.	B	70-71
3.	A	70
4.	A	70
5.	C	70
6.	A	70
7.	D	70-72
8.	B	72
9.	E	72
10.	A	72

Chapter 7 – Capsule Adhesive Anchor Systems

Question	Correct Answer	Reference
1.	A	77
2.	C	77
3.	B	77-79, 88
4.	A	88
5.	A	88-90
6.	A	88
7.	D	89-90
8.	B	79, 87
9.	A	79
10.	B	88

Chapter 8 – Installation Problems

Question	Correct Answer	Reference
1.	A	94
2.	D	94
3.	B	94
4.	D	94
5.	D	95
6.	A	96
7.	A	95
8.	C	95
9.	B	95
10.	A	95

APPENDIX A

CONCRETE INTERNATIONAL ARTICLES

Wollmershauser, Richard E., and Mattis, Lee, "Adhesive Anchor Installation and Inspection," *Concrete International*, V. 30, No. 12, Dec. 2008, pp.36-40.

Grosser, Philipp; Fuchs, Werner; and Eligehausen, Rolf, "A Field Study of Adhesive Anchor Installations, " Concrete International, V. 33, No. 1, Jan. 2011, pp. 57-63.

Adhesive Anchor Installation and Inspection

Understanding the requirements to ensure proper performance

BY RICHARD E. WOLLMERSHAUSER AND LEE MATTIS

Post-installed adhesive anchors, also known as epoxy anchors, chemical anchors, bonded anchors, and adhesive-bonded anchors, have been successfully used to make connections to concrete structures for many years. With the failure of adhesive anchors in the Boston I-90 Tunnel Project, commonly called the Big Dig, the use of these types of anchors has been called into question. This article reiterates the proper installation and associated inspections necessary for adhesive anchors to achieve the desired performance.

WHAT ARE ADHESIVE ANCHORS?

Adhesive anchors in concrete or masonry derive their resistance to applied tension load by adhesion or bond. The adhesive, available in both cartridge and capsule configurations, consists of two essential components: a resin and a hardener.

When cartridges are used, the two components are contained in separate parallel tubes connected to a manifold that proportions the materials in the proper ratio and mixes them together. The cartridge tool forces the materials out of the tubes, through the manifold and mixing nozzle, and into the drilled hole. The mixing nozzle ensures that the components are well mixed so the adhesive resin is activated by the hardener.

When an adhesive capsule is used, the resin and hardener are kept separate but are contained within a single glass or foil capsule. After inserting the entire capsule into the drilled hole, the anchor element, usually a threaded rod, is inserted into the predrilled hole with a rotational motion using a rotary drill. The rotary motion of the anchor breaks the capsule, causing the resin and

hardener to mix and initiating the chemical reaction that hardens the adhesive.

Adhesive anchors are available in a variety of chemistries, each with its own specific characteristics and capacities. The adhesive materials include epoxies with many different formulations, acrylates, vinyl esters, polyesters, and hybrid mortars. The specifier, installer, and end user should become familiar with the requirements of each specific application to ensure the selected anchor and adhesive materials are appropriate for the given application.

SELECTION OF ADHESIVE ANCHORS

Selection of the appropriate adhesive anchor system requires an understanding of the loads to be resisted. Not only the direction of the load (tension, shear, or a combination of tension and shear), but also the duration (sustained or short-term loads) and the source (gravity, wind, or seismic). Proper selection also requires matching the adhesive material to the environment of the application. This includes consideration of such factors as the expected ambient environment, elevated temperatures, and protection from adverse weather.

PREQUALIFICATION OF ADHESIVE ANCHORS

Prior to 1995, there were no written criteria for testing and qualifying adhesive anchors. The International Conference of Building Officials Evaluation Service (ICBO ES) recognized adhesive anchors through AC01, "Acceptance Criteria for Expansion Anchors in Concrete and Masonry Elements,"[1] as a basic test protocol and acceptance criteria with some additional special requirements applicable to adhesive anchors. Recognizing

A–2

the need for stand-alone adhesive anchor criteria, ICBO ES requested recommendations from the anchor industry. In 1994, an ad hoc committee began work on draft acceptance criteria for adhesive anchors. The committee developed a set of testing protocols that could be used to prequalify adhesive anchors for use with the design provisions contained in the Uniform Building Code (UBC).[2]

In January 1995, ICBO ES adopted AC58, "Acceptance Criteria for Adhesive-Bonded Anchors in Concrete and Masonry."[3] This document contained the testing methods (based on ASTM E1512[4]) and acceptance criteria for evaluating and qualifying adhesive anchors for structural use in accordance with the provisions of the UBC. The Concrete Anchor Manufacturers' Association (CAMA) was subsequently established by members of the ad hoc committee to assist with future criteria and code development.

Since the merger of the three major building code organizations into the International Code Council (ICC) and the subsequent creation of the ICC Evaluation Service (ICC-ES), AC58 has remained in use. With the publication and adoption of the 2003 International Building Code (IBC 2003),[5] however, new qualification criteria were required to address the requirements of ACI 318-02,[6] Appendix D: Anchoring to Concrete, which is referenced in the IBC. Although ACI 318-02 excluded adhesive anchors from the design procedures, CAMA provided recommendations for a new design procedure and acceptance criteria in AC308, "Acceptance Criteria for Post-Installed Adhesive Anchors in Concrete Elements,"[7] to address the requirements in Appendix D. AC308 was adopted by ICC-ES in June 2005 and now supersedes AC58 for concrete installations. AC308 contains acceptance criteria for adhesive anchors used in both cracked (tension zone) and uncracked concrete in accordance with the provisions of the IBC.

After testing by an independent third party according to the protocols in AC308, ICC-ES evaluates the testing and issues an Evaluation Service Report (ESR) that contains all the information necessary to properly use the adhesive anchor system including technical data, installation procedures, the category of the anchor system for design, and any limitations on the adhesive anchor system.

INSTALLATION REQUIREMENTS

Once the correct adhesive anchor system has been selected, installation is the next critical aspect to be considered. As part of the prequalification procedure, anchors are tested after they're installed according to the manufacturer's published installation procedures. It's of great importance to follow these instructions to obtain the published anchor capacities and performance characteristics. Installation instructions specify the

drilling method, hole cleaning procedure, how to install the adhesive and the metal anchor in the drilled hole, and the care to be taken until the adhesive has cured. Let's look at each of these aspects.

Drilling method

The usual method for drilling concrete is to use a rotary-hammer drill with a carbide drill bit of a specified diameter, creating a roughly cylindrical hole. Prequalification tests are typically based on such a drilling system.

Other drilling methods sometimes used include rock drills and diamond coring. Rock drills usually produce holes with walls that are somewhat rougher than those drilled with rotary hammer drills, whereas diamond core bits usually produce holes with walls that are somewhat smoother. Both of these drilling methods affect the capacity of adhesive anchors—rock drilling generally increases the capacity, whereas core drilling generally decreases the capacity. The user should check whether the anchor capacity is affected if other than rotary hammer drilling is used. Because diamond core drilling is

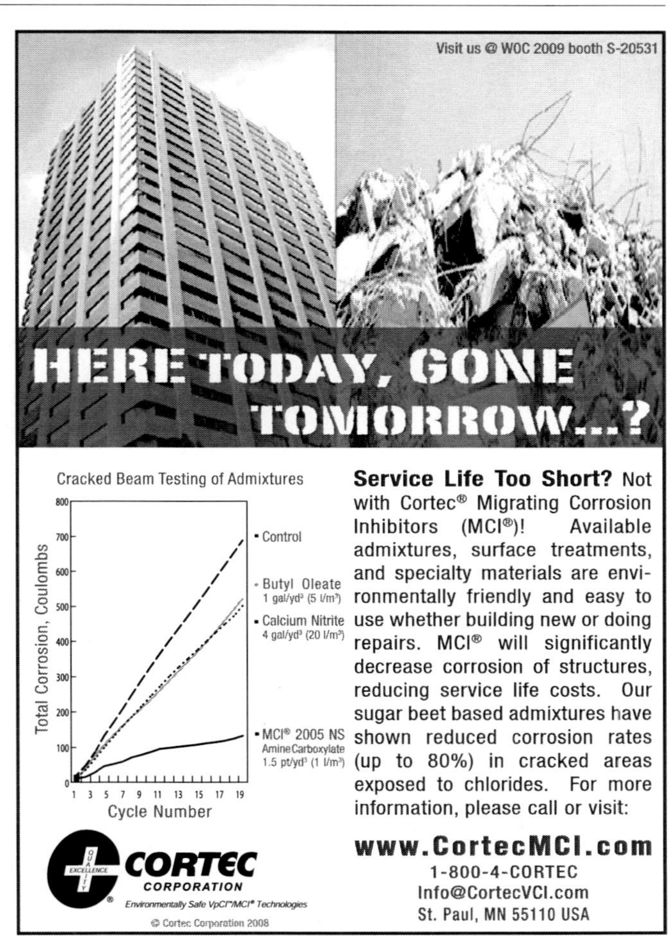

CIRCLE READER CARD #15

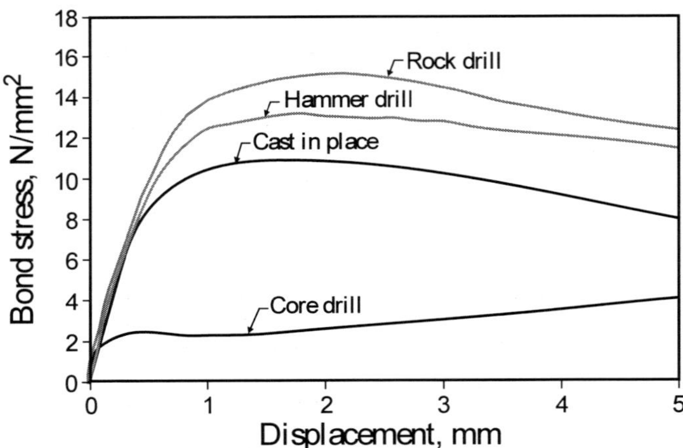

Fig. 1: Bond stress-displacement curves for 20 mm (0.8 in.) diameter adhesive anchors installed in holes made by various drilling methods. Anchors in core drilled holes had 300 mm (11.8 in.) embedment in 50 MPa (7250 psi) concrete. All other anchors had 200 mm (7.9 in.) embedment in 20 MPa (2900 psi) concrete (based on Reference 8) (1 in. = 25.4 mm, 1000 lb/in.² = 6.89 N/mm²)

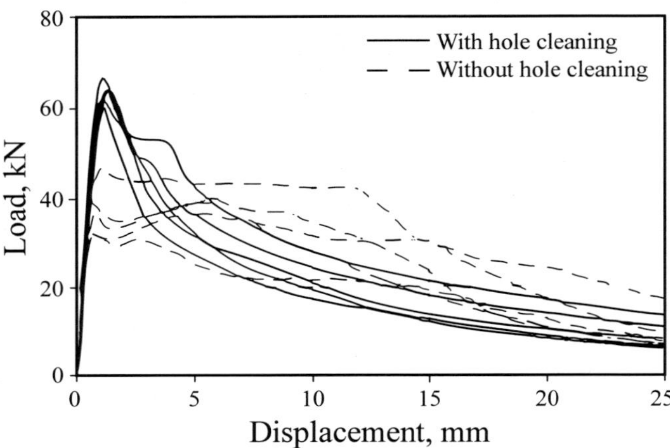

Fig. 2: Load-displacement curves for 12 mm (0.5 in.) diameter anchors with 110 mm (4.3 in.) effective embedment depth in 24 MPa (3500 psi) concrete bonded using an adhesive that is sensitive to hole cleaning (based on Reference 8) (1 in. = 25.4 mm, 1000 lb = 4.45 kN)

usually accomplished using water, the method creates a slurry. This slurry is not removed while drilling, so the residual material must be flushed from the hole before it dries to prevent it from affecting the anchor tensile capacity. Figure 1 illustrates the reduction in bond strength that may be obtained from different drilling methods compared to a cast-in reinforcing bar. The adhesive used was a low-adhesion resin system.

Some anchor systems are sensitive to the size of the annular gap between the anchor rod and the wall, while others are not. The correct bit type and diameter must be used in accordance with the anchor manufacturer's instructions.

Because the performance of the anchor depends on a specified embedment into the concrete, the hole must be drilled to the depth recommended by the manufacturer.

Hole cleaning

Proper hole cleaning is critical to almost all adhesive anchor installations and the lack thereof is frequently a major source of poor adhesive anchor performance. Figure 2 demonstrates the influence of lack of hole cleaning on the load-displacement curves of adhesive anchors. The specified procedures must be followed for the adhesive to properly adhere to the concrete. The usual method involves brushing with a stiff metal or nylon brush and blowing out the hole with sufficient compressed air. This must be done in the sequence and manner that best removes the debris and dust particles from the hole. Figure 3 illustrates the influence of the degree of hole cleaning on bond strength for three different injection systems.

Any standing water from rain or construction activity should be thoroughly removed and the hole should be dry during anchor installation, unless the anchor system has been qualified for installation under damp or wet conditions. Figures 4 and 5 show the possible effects of installing anchors in wet concrete using a resin that is sensitive to moisture. The correct installation procedures will vary by product but will be specified in the manufacturer's literature and the ESR.

Installation of the adhesive and anchor

For cartridge adhesive systems, the manufacturer's complete system should be used. When a new cartridge is used, manufacturers universally recommend that some adhesive be dispensed and wasted on a disposable surface until a consistent adhesive color is obtained. This ensures that the adhesive materials are properly mixed and will set and harden properly. If this procedure is not followed, the first anchor installations with the new cartridge may not achieve full performance. In some cases, the anchors can be removed by hand because the adhesive is not adequately mixed. After hole cleaning, this is one of the most common and serious errors that occur with cartridge systems. Anecdotes are told about installations where every tenth (or some other multiple) anchor could be easily removed by hand, corresponding to the use of a new cartridge.

The mixing nozzle is inserted into the bottom of the hole and slowly withdrawn as the adhesive is pumped into the hole while avoiding the introduction of air voids into the adhesive mass. It's important to fill the hole with enough adhesive so the annular space is completely filled when the anchor rod is installed. The manufacturer's installation instructions typically state how much adhesive should be dispensed into the hole to completely fill the annular space. Insertion of the anchor element should be

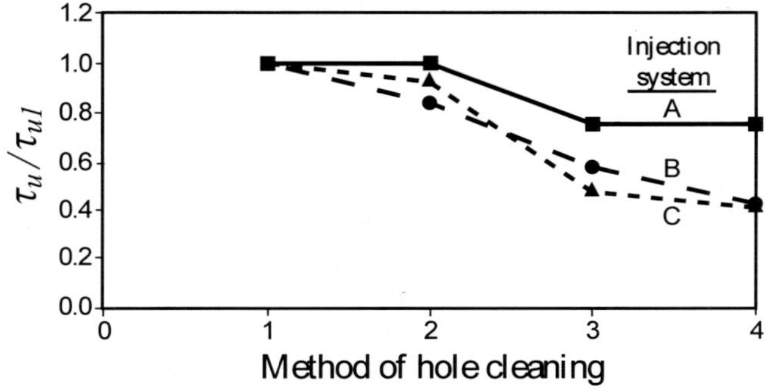

Method of hole cleaning	
1	2 x blowing 2 x brushing 2 x blowing
2	1 x blowing 1 x brushing 1 x blowing
3	2 x blowing
4	No cleaning (drilling machine retracted 3 times)

Fig. 3: Influence of the intensity of hole cleaning on the bond strength (τ_u) of 12 mm (0.5 in.) diameter injection anchors in dry concrete (τ_{u1} = bond strength using cleaning method 1) (based on Reference 8)

with a twisting motion to ensure that the adhesive flows into the threads of a rod or around the deformations of a reinforcing bar. It's very important that the anchor element is inserted to the bottom of the hole so the required embedment is achieved. The presence of air voids, incomplete hole filling, and short embedment reduce the area of adhesive bond to the anchor rod or the wall of the hole and are detrimental to adhesive anchor performance.

Capsule systems are much less sensitive to installation procedures. The manufacturer's installation procedure, however, should be followed. The capsule is inserted into the drilled hole. The anchor, usually a threaded rod, is inserted to the full depth of the hole with a rotary hammer drill, mixing the adhesive contents of the capsule, as well as scouring the wall of the hole. The depth of the hole is specified by the manufacturer and corresponds to the capsule diameter and desired embedment length. The total length of the anchor rod depends on the embedded length and the amount of the anchor that will protrude above the concrete. Once the insertion of the anchor is completed, the anchor should not be touched until the adhesive has set.

Curing

After the anchor is inserted into the adhesive, it should not be disturbed until the adhesive has cured. Set times are listed in the ESR or the manufacturer's instructions. Curing takes more time at lower temperatures and less time at higher temperatures. Some adhesives can be loaded in tension within an hour of installation, while others may require as long as 24 hours. Refer to the ESR or the manufacturer's literature for product-specific information.

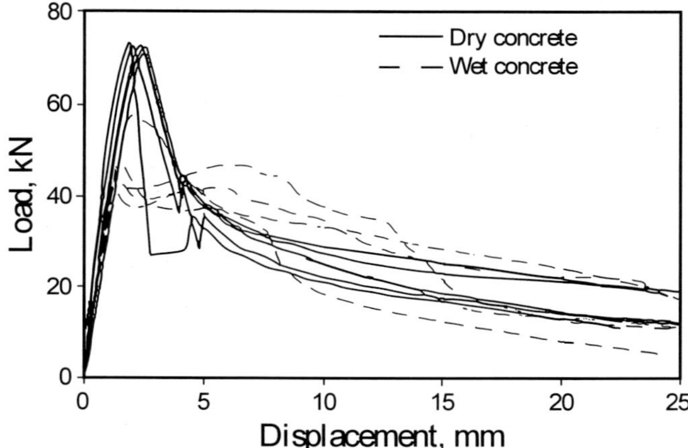

Fig. 4: Load-displacement curves for 12 mm (0.5 in.) diameter anchors with 110 mm (4.3 in.) effective embedment depth in 24 MPa (3500 psi) concrete. Anchors installed in thoroughly cleaned holes and bonded using an adhesive that is sensitive to moisture in concrete (based on Reference 8) (1 in. = 25.4 mm, 1000 lb = 4.45 kN)

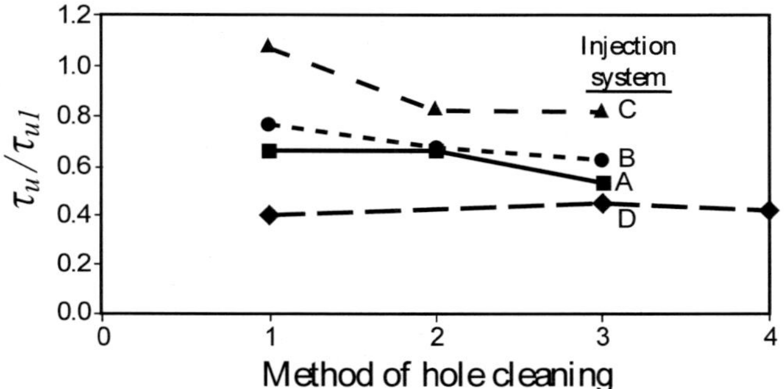

Method of hole cleaning	
1	2 x blowing 2 x brushing 2 x blowing
2	1 x blowing 1 x brushing 1 x blowing
3	2 x blowing
4	No cleaning (drilling machine retracted 3 times)

Fig. 5: Influence of the intensity of hole cleaning on the bond strength (τ_u) of 12 mm (0.5 in.) diameter injection anchors in wet concrete (τ_{u1} = bond strength using cleaning method 1 in dry concrete) (based on Reference 8)

SPECIAL INSPECTION

Installation of adhesive anchors under the IBC requires special inspection to ensure that the installation is correctly performed in accordance with the design and the ESR requirements. Special inspection is continuous observation of construction activities requiring unique expertise or where additional assurance of quality is deemed necessary. The requirements are found in IBC Sections 1702, 1704, and 1901.5. The owner or the licensed design professional, acting as the owner's agent, must employ the special inspector. The special inspector must be approved by the building official and has a responsibility to the building official in the performance of special inspection.

Special inspection is mandated by the ESRs for all adhesive anchor systems. This is due to the correct perception by ICC-ES that proper installation is critical to achieving the desired performance. The special inspector must verify that the installation is in accordance with the requirements of the approved plans, ESR, and manufacturer's instructions. This generally means verifying the location of the anchor, any edge distance and spacing requirements, drilling equipment, drill bit type and size, hole depth, hole cleaning procedures, anchor type and material, size, embedment, and installation procedure including checking adhesive expiration date and proper dispensing meet the specification.

Although special inspection is defined as continuous observation of the construction activity, there is a provision in the IBC for periodic inspection with the approval of the building official. Because critical requirements can be verified at certain phases of the installation, the installation of adhesive anchors can effectively be inspected periodically. For instance, the special inspector can verify appropriate requirements after the holes have been drilled and cleaned but prior to installation of the adhesive and anchors. The inspection of the adhesive dispensing and anchor installation can be done immediately after the verification of hole drilling with the special inspector present. In any case, the special inspector must be able to directly witness and verify all aspects of the installation.

Proof loading alone is not recognized as meeting special inspection requirements. While proof loading may be specified as a supplemental requirement to special inspection, visual inspection of the anchor installation procedures must still be provided because it's not possible to verify embedment and other important installation requirements after the anchors have been installed. From a practical standpoint, it's not always possible to load an anchor to a level that will adequately stress the adhesive due to limitations of the anchor material properties, such as the steel strength. For instance, a proof load of 50% of the adhesive bond strength is considered a reasonable maximum load that will not damage the adhesive bond of a properly installed anchor. This load level, however, may not be possible if a mild steel anchor rod is used because it would load the rod close to or above its yield strength. Nevertheless, proof loading can be an effective incentive to promote good installation practices.

For an in-depth article about special inspection for both mechanical and adhesive post-installed anchors, refer to the CAMA Web site, **www.concreteanchors.org**, where a comprehensive paper[9] on this subject can be found under the "PUBLICATIONS" link.

References

1. AC01, "Acceptance Criteria for Expansion Anchors in Concrete and Masonry Elements," ICC Evaluation Service, Inc., Jan. 1993, 7 pp.

2. Uniform Building Code 1997, "Volume 2—Structural Engineering Design Provisions," International Conference of Building Officials, Whittier, CA, 1997, 492 pp.

3. AC58, "Acceptance Criteria for Adhesive-Bonded Anchors in Concrete and Masonry," ICC Evaluation Service, Inc., Jan. 1995, 28 pp.

4. ASTM E1512-01(2007), "Standard Test Methods for Testing Bond Performance of Bonded Anchors," ASTM International, West Conshohocken, PA, 2007, 5 pp.

5. "International Building Code 2003," International Code Council, Falls Church, VA, Dec. 2002, 660 pp.

6. ACI Committee 318, "Building Code Requirements for Structural Concrete (ACI 318-02) and Commentary (318R-02)," American Concrete Institute, Farmington Hills, MI, 2002, 443 pp.

7. AC308, "Acceptance Criteria for Post-Installed Adhesive Anchors in Concrete Elements," ICC Evaluation Service, Inc., June 2005, 122 pp.

8. Eligehausen, R.; Mallée, R.; and Silva, J., *Anchorage in Concrete Construction*, Ernst & Sohn, Berlin, 2006, pp. 187-189.

9. Mattis, L., "Special Inspection Guidelines for Expansion and Adhesive Anchors," Concrete Anchor Manufacturers Association, St. Charles, MO, 2002, 7 pp. (available at www.concreteanchors.org/pubs.htm).

Selected for reader interest by the editors.

Richard E. Wollmershauser, FACI, is a consultant from Tulsa, OK. He is the Chair of ACI Committee 503, Adhesives for Concrete, and a member, past Chair, and past Secretary of 355, Anchorage to Concrete. He also currently chairs ASTM Subcommittee E06.13, Performance of Connections in Construction, and is a member of several other anchor-related committees. He was previously Director of Technical Services for Hilti, Inc., and has more than 25 years of experience in the field of anchor development, testing, and qualification.

ACI member **Lee Mattis** is Principal Engineer of CEL Consulting, Oakland, CA, where he supervises the testing of concrete anchors and other construction materials and engineering activities. He has more than 30 years of experience in the field of concrete anchor testing, inspection, and design. He is a member of ACI Committee 355, Anchorage to Concrete, as well as several other U.S. and international technical committees.

A Field Study of Adhesive Anchor Installations

Theory and practice

BY PHILIPP GROSSER, WERNER FUCHS, AND ROLF ELIGEHAUSEN

For every fastening application, the starting point is choosing the correct product. This requires understanding the loads to be resisted and following a defined design method. It also requires establishing the exact location of the fixing point, anticipating the environmental conditions during installation and service, and knowing the installation procedures for the fastener or anchor.

For adhesive anchors, the designer must select a product that is adequate for the conditions in service and that has been evaluated for compliance with code requirements.

Once an appropriate system has been selected and designed, the installation process has to be considered. To ensure correct installation, the installer must follow the manufacturer's instructions precisely. Furthermore, the installer must follow the requirements of the relevant evaluation service report (ESR)—refer to the sidebar. Detailed information on the installation and inspection of adhesive anchors is given by Wollmershauser and Mattis,[1] and information on factors influencing the behavior of adhesive anchors can be found in Reference 2.

In theory, the knowledge is available to ensure reliable fastenings with adhesive anchors and to give designers and installers confidence and flexibility in myriad applications. With the failure of adhesive anchors in Boston, MA, however, the installation and use of these types of anchors has been called into question. To figure out what can be improved with regard to the use of adhesive anchors, adhesive anchor installations with injection systems were monitored on 23 job sites in five locations scattered over the U.S. Critical aspects were examined to determine gaps between actual and recommended practice and to come up with proposals to improve installation practices.

FIELD RESEARCH PROJECT

The installation of adhesive anchor systems was investigated at construction sites in California, Florida, Illinois, New York, and Pennsylvania. In total, 23 sites were visited, 26 applications were monitored, and 31 installers were interviewed (Table 1). Thirteen different adhesive systems (either epoxy-based or hybrid mortars) were installed. Nine of these products had an ESR. The steel anchors were continuously threaded steel rods or deformed steel reinforcing bars. Anchors were used for strengthening bridge structures, seismic retrofits, connecting structural elements to structural walls, anchoring steel elements to existing concrete members, anchoring pavement dowels, installing hurricane protection, or anchoring façade elements. The anchors were installed downward in 13 applications, horizontally in 11 applications, and overhead in two applications.

All information relevant to the installation was monitored on site by the first author, and the detailed findings were recorded in a protocol. Installers were interviewed to determine their professional and educational backgrounds, training and experience levels with post-installed anchors, general anchor installation experience level, and general opinions concerning the pros and cons in regard to the installation of adhesive-bonded anchors.

TABLE 1:

DETAILS OF THE SURVEY

Location	Illinois	Florida	California	Pennsylvania	New York	Total
Job sites	6	7	5	2	3	23
Applications	7	8	5	3	3	26
Surveys	8	6	8	5	4	31

General observations

Figure 1 summarizes the general observations made at the construction sites. All manufacturers' product installation instructions (MPIIs) were analyzed to evaluate whether the required information to ensure a correct installation was given. Nine out of the 13 products (five with ESRs and four without ESRs) did not deliver all information that's necessary for the correct installation of adhesive anchors (Fig. 1(a)).

Some MPIIs were ambiguous, some had incomplete instructions, and others had pictograms that did not match the information in the text. With some products, the instructions were only printed on the cartridge and with a very small font—this information couldn't be read after the cartridge was placed in the dispenser. Whereas most systems required that the adhesive be stored in a dry, relatively dark, and well-ventilated area at temperatures within a specified range, it was found that some MPIIs

gave no information about storage requirements. Some MPIIs provided no information about the diameter of the drill bit relative to the anchor size, and none of the MPIIs indicated that a depth stop was required. Table 2 contains a brief summary on the various ways the MPIIs statements failed to provide adequate detail.

In most of the applications, the MPIIs for the adhesive systems were available on site, but many installers didn't refer to the instructions. In one application, the product was past its expiration date, but it was used anyway. The adhesives were found to be stored in various places that ran counter to the storage requirements (Fig. 1(b)).

Figure 2 provides specific observations made during anchor installations, including borehole drilling, cleaning, and condition; adhesive dispensing and injection; and curing time. These observations are discussed in detail in the following sections.

Borehole

Depending on the product, the anchor capacity might be reduced if the borehole has the wrong depth, location (Fig. 3 and 4), or diameter (Fig. 5). Installers should use a drill bit meeting the tolerances stated in the MPIIs, but it was observed that an incorrect drill bit was used in many applications. In 23% of the applications, the drill bit was too large (the maximum deviation was 33%), and in 15% of the applications, the drill bit was too small. In one extreme case, the drill bit was about the same size as the nominal anchor rod size (Fig. 5). As indicated previously, some MPIIs didn't define the proper bit size, so the use of improper drill bits isn't entirely surprising.

In most of the installations, a depth stop (depth gauge) wasn't used (Fig. 6). While none of the MPIIs included a requirement to use a depth stop, all stated that the borehole must be drilled to the correct size and depth; this is only possible when using adequate accessories and aids.

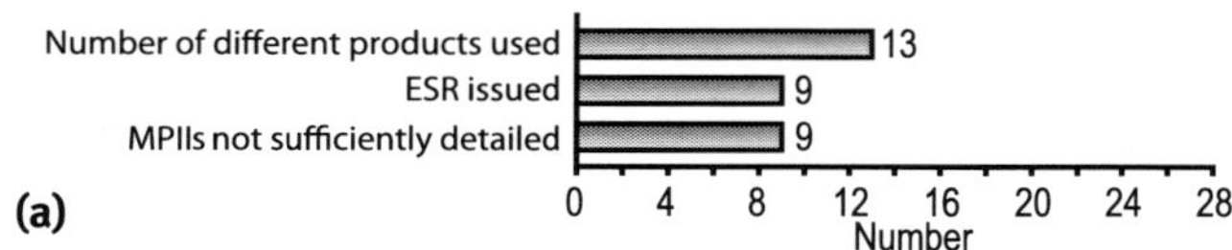

(a)

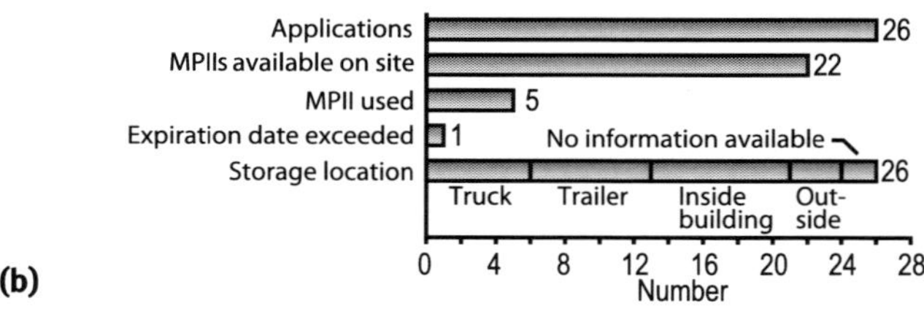

(b)

Fig. 1: General observations made on job sites: (a) number of different products used and information regarding ESR and MPIIs; and (b) monitored applications and observations regarding use of MPIIs, age of adhesive, and storage of adhesive

Product	1	2[*]	3	4[*]	5[*]	6	7[*]	8[*]	9
Storage	X	X	X						X
Temperature of cartridge during installation	X	X	X			X			X
Temperature of base material during installation					X				
Drilling method	X		X	X	X	X	X	X	X
Drill bit size				X	X	X	X		X
Condition of borehole (dry, wet, water-filled)	X		X	X	X	X		X	
Cleaning procedure						X			X
Gel time as a function of temperature			X						
Installation torque	X		X		X	X		X	X

[*]Product with ESR

All product instructions required blowing out the drilled hole with either a hand pump or compressed air (Fig. 7). All instructions also required additional brushing of the borehole. In general, however, the correct tools required for cleaning the boreholes were not available on site (Fig. 8). The boreholes were properly cleaned in only a small number of installations (Fig. 2). Most adhesive systems require that the boreholes are free of standing water, dust, debris, ice, grease, oil, or other foreign materials. In two applications, however, the anchor was installed in a water-filled hole, whereas the ESR stated that the product was not suitable for this installation condition.

Adhesive installation and curing

The MPIIs require the installer to dispose of the initial mixture of hardener and resin pushed through the mixing nozzle (the mixture dispensed into the borehole should have a uniform color). This procedure ensures that the correct ratio of hardener and resin is used. These requirements were often not met (Fig. 2). Many installers reused the mixing nozzle from a previous cartridge. The old adhesive within the nozzle was discarded, but the first part of the adhesive from the new cartridge was dispensed into the borehole. It can be assumed that the actual ratio between hardener and resin for this material is different from the mixing ratio required for full conversion, so this practice likely reduces the capacity of the installed bonded anchor.

To fill the hole with adhesive (and avoid air pockets), the adhesive should be injected from the bottom of the hole and the nozzle should be withdrawn as filling progresses. In some applications, however, the adhesive was not injected from the bottom of the hole. In these cases, the borehole wall and steel part may not be fully covered by the adhesive and voids may be present in the hardened mortar. Both effects yield a significant reduction in capacity and durability. Furthermore, depending on the product, oxygen inhibition might interrupt the curing process of the adhesive and therefore negatively influence the anchor capacity.

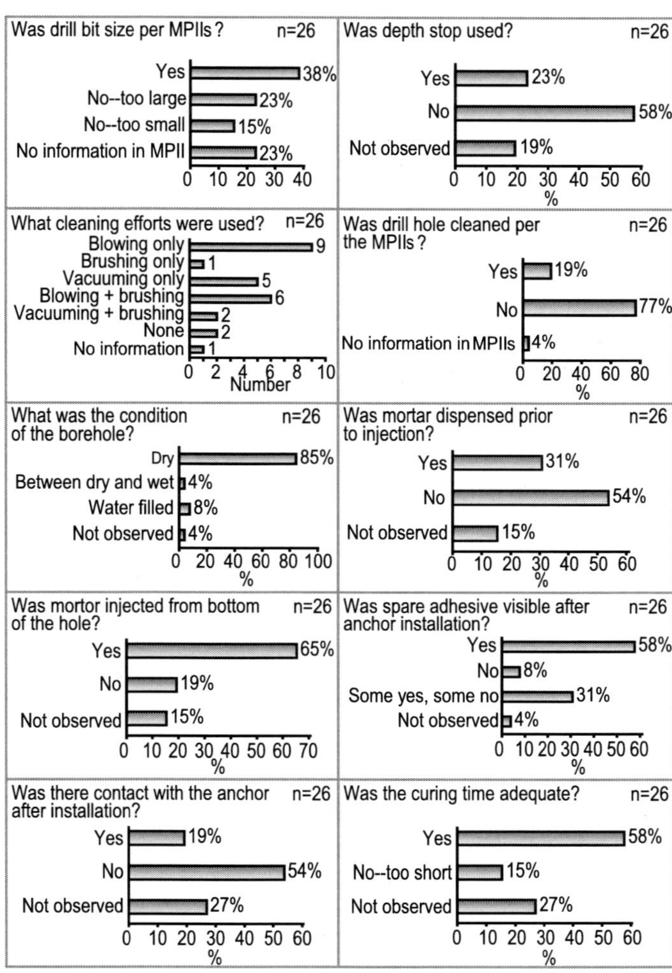

Fig. 2: Results of the field study of adhesive anchor installations. (n = number of observed applications)

After inserting the anchor rod or bar, there must be spare adhesive visible all around the mouth of the hole to ensure that the borehole is fully filled with mortar. Spare adhesive wasn't visible, however, in most of the applications. Because of the missing mortar, the anchor rod is not

Fig. 3: To avoid interference problems, reinforcing bars must be located before drilling and provisions must be made in the details to allow adjustment of anchor locations. Otherwise, bars would be drilled through reinforcement and affect the capacity of the structure. Here, the installer attempted to clear initially undetected bars by inclining the boreholes. The long extensions on two of the threaded rods show that the attempt was a failure

Fig. 4: Details must set limits on how anchor locations are adjusted. In this installation, a second borehole (as well as a new hole in the base plate) had to be drilled because the drill bit contacted a reinforcing bar at the original location. The anchorage capacity will be reduced, however, because the resulting anchor spacing is too small

Fig. 5: After the boreholes were filled with adhesive, these threaded rods were hammered into place because the hole was too small. The size of the drill bit required for a given anchor diameter should be clearly communicated in the MPIIs

Fig. 6: The correct method for creating a borehole. The installer is drilling through a template using a hammer drill with a depth stop

A–10

bonded to the borehole wall over its full length and its capacity will be reduced.

Also, after inserting the anchor into the adhesive, each system requires a curing time that depends on the temperature of the base material. During this time, it's necessary to ensure that the rod is not disturbed or loaded. Figure 9, for example, shows a correctly installed overhead application. Early contact or premature loading could partially destroy the bond between the concrete and mortar. In some applications, however, rod contact was observed and loading was applied during the curing time.

Special inspection

Continuous or periodic inspection for anchor installations was only found in the five applications in California and one of the eight applications in Florida (Fig. 10). No special inspections were observed in the seven applications in Illinois, the three applications in Pennsylvania, or the three applications in New York. Whereas all boreholes were cleaned properly and all anchors were installed correctly in the California applications, this was not the case in Florida. The authors highly recommend that the engineer in charge of a project reviews the performance of the inspectors observing anchor installations.

Installers

After monitoring the installations on site, we had 31 installers fill out a questionnaire. The completed questionnaire gave information on adhesive anchor installations on other sites and the educational background of the installers.

Fig. 7: The correct method for cleaning a hole. The installer is blowing dust out of the hole according to the relevant MPIIs using compressed air directed through a nozzle

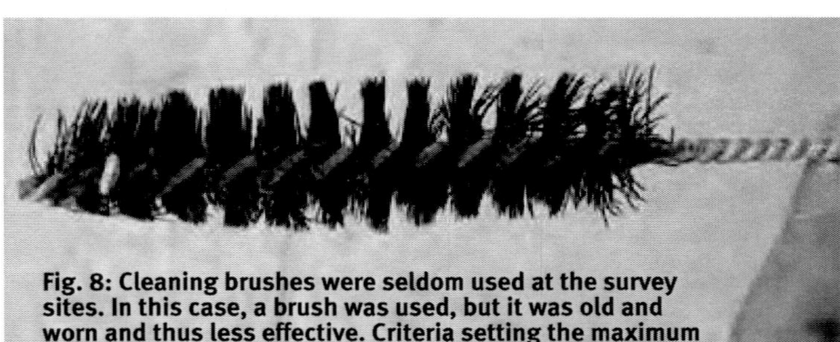
Fig. 8: Cleaning brushes were seldom used at the survey sites. In this case, a brush was used, but it was old and worn and thus less effective. Criteria setting the maximum acceptable wear were not provided in the MPIIs available at the job site

Fig. 9: This overhead installation was completed correctly. Excess adhesive around the anchor indicates the borehole was properly filled, and the installer used wedges to hold the anchor in place during curing

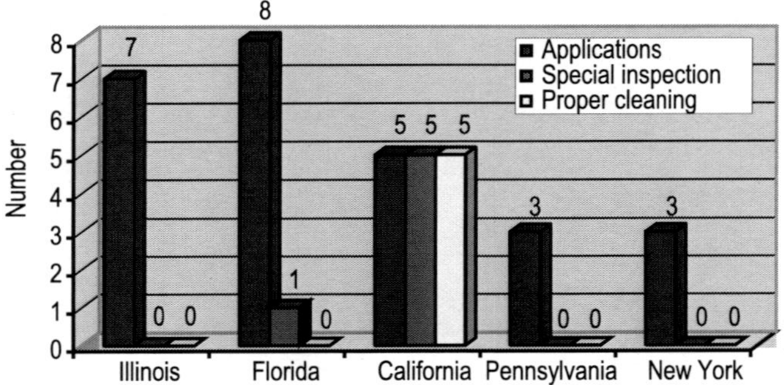

Fig. 10: Influence of special inspection of anchor installations. Proper borehole cleaning was only observed in California, a state with a long history of special inspection dating back to the UBC

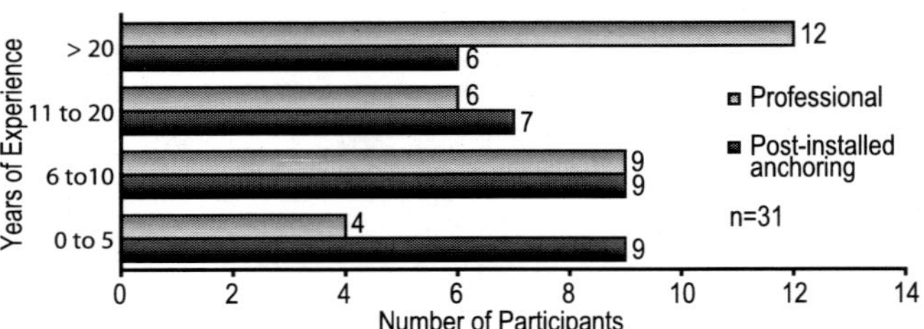

Fig. 11: The installers had significant experience in construction and with post-installed anchor installations. (n = number of installers interviewed)

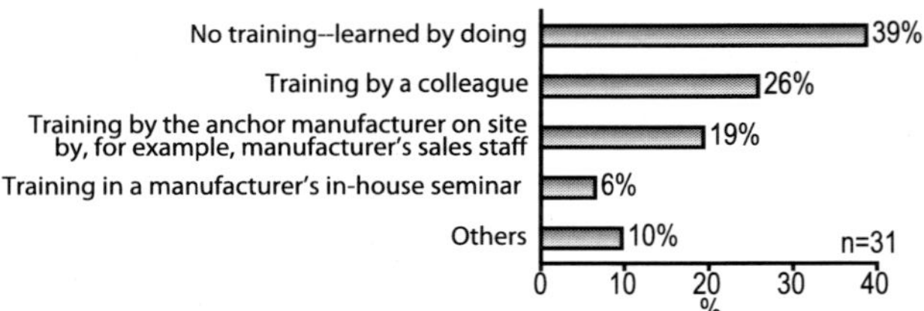

Fig. 12: Training of installers. Although the installers wanted to do a good job, their training was inadequate. (n = number of installers interviewed)

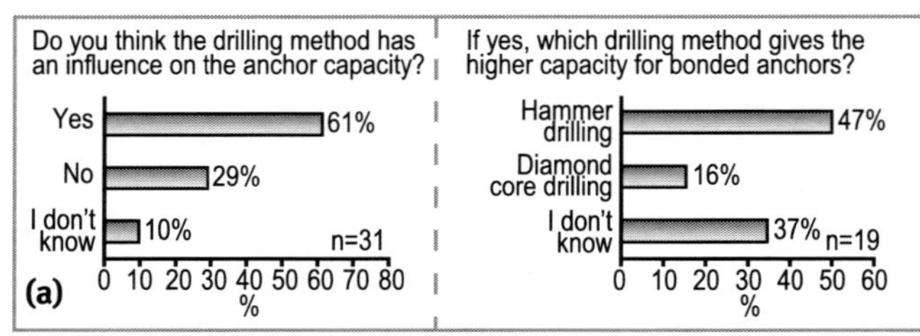

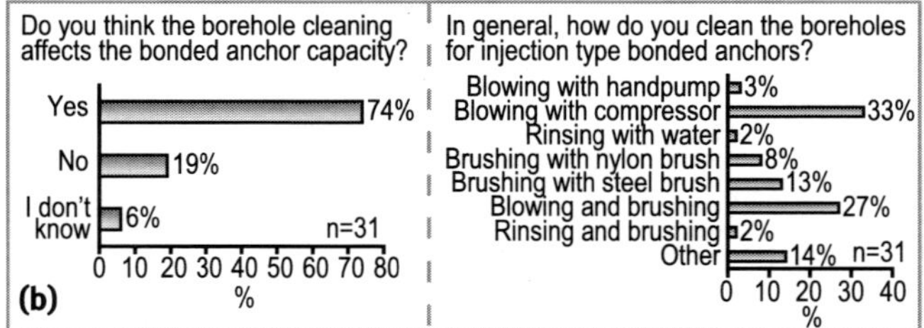

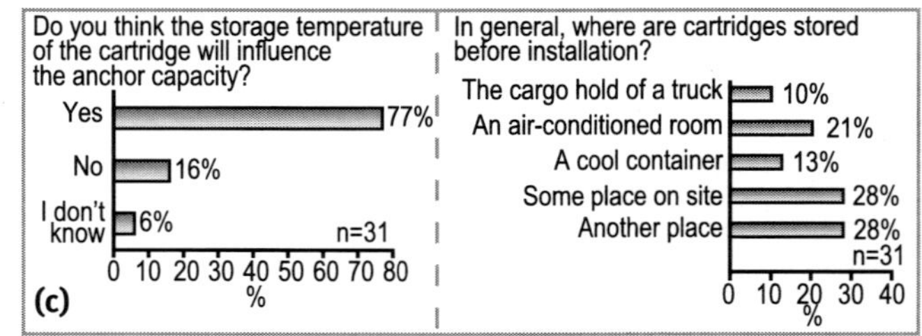

Fig. 13: Many installers were not aware of the influences of significant factors on anchor capacity: (a) Although about 60% of the installers thought that the drilling method was significant, only half of these knew which drilling method provided higher capacity; (b) Although almost three out of four installers thought that the borehole cleaning method has an effect, less than a third were using blowing and brushing; and (c) Although over three out of four installers thought that storage temperature has an influence, only about a third were keeping the cartridges cool during storage. (n = number of installers interviewed)

Almost all of the installers had technical backgrounds— the majority had been carpenters, millwrights, ironworkers, or general construction workers. Most of the installers had installed anchors for several years. Figure 11 summarizes the years of experience in their trades and in post-installed anchoring technology.

It was evident that the majority of installers wanted to do a very good job, tried to handle the installation process carefully, and were eager to improve their knowledge in fastening technology and installation procedures. In general, however, they were not informed about improvements in knowledge and procedures. The evaluation of the questions with regard to installation training is given in Fig. 12. Sixty-five percent of all installers stated that they gained their knowledge in post-installed anchoring technology either from "learning by doing" or from a colleague. One-quarter answered that they were trained by a manufacturer on site or in an in-house seminar, and only 10% answered that they got their knowledge from other sources such as schools.

Installers were also asked for their opinions regarding the influence of their practices on the capacity of adhesive anchors. The results are shown in Fig. 13. Whereas most of the installers interviewed knew that the drilling method has an influence on the anchor capacity (Fig. 13(a)), only half of them knew that the rougher borehole wall provided by hammer drilling gives a higher tensile capacity than the smooth borehole wall provided by diamond core drilling. Some of the installers did not know about the influence of borehole cleaning (Fig. 13(b)) on capacity. Note that correct borehole cleaning was mainly found in California, which might be attributed to the effective special inspection observed at those job sites. Most of the installers answered that the storage temperature has an influence on the anchor capacity; however, many stated that they store the adhesive on site or somewhere else, where storage temperature cannot be controlled (Fig. 13(c)).

A–12

FURTHER ACTIONS

The critical factors influencing the load-bearing behavior of adhesive-bonded anchors are well known. The findings summarized herein, however, demonstrate that there might be room for improvement to ensure high installation quality.

Training and certification

This study indicates that installers are generally very eager to do a good job, but much of the information available in ESRs isn't being used. Installers either haven't been properly trained, or they have been given inadequate installation instructions. There is an urgent need to improve the installation process for adhesive anchors on site, so installers need more detailed education in installing these types of anchors. Therefore, ACI Committee C601-A, Adhesive Anchor Installer, is working jointly with the Concrete Reinforcing Steel Institute (CRSI) on a new certification program. The development of the ACI-CRSI Adhesive Anchor Installation Certification Program has been on an accelerated schedule, and its rollout is scheduled for early 2011.

Other measures

In addition to proper training, other measures might help improve the quality of adhesive anchor installations. All parties have an incentive to demand that the manufacturer's product installation instructions are written and illustrated in a uniform, clear, and unmistakable way, and that those instructions are readily available to the installer and inspector.

For critical applications, effective special inspection seems to be necessary to ensure that anchors are installed correctly. Training of inspectors would help ensure effective special inspection.

Proof-loading can be effective in improving installations because it allows detection of installations with significant reductions in capacity. The contractor will have a strong incentive to ensure correct anchor installation and avoid problems that will develop if proof-loading criteria are not met.

Acknowledgment

We'd like to express special thanks to Powers Fasteners, Inc., for its support of the field research project.

References

1. Wollmershauser, R., and Mattis, L., "Adhesive Anchor Installation and Inspection," *Concrete International*, V. 30, No. 12, Dec. 2008, pp. 36-40.

2. Eligehausen, R.; Mallée, R.; and Silva, J.F., *Anchorage in Concrete Construction*, Ernst & Sohn, Berlin, Germany, 2006, pp. 181-210.

3. *International Building Code*, International Code Council, Inc., Country Club Hills, IL, 2009, 676 pp.

4. AC308, *Acceptance Criteria for Post-Installed Adhesive Anchors in Concrete Elements*, ICC Evaluation Service, Whittier, CA, 2009, 122 pp.

5. Uniform Building Code, V. 2, Structural Engineering Design Provisions, 1997 edition, International Council of Building Officials, Whittier, CA, 1997, 492 pp.

Received and reviewed under Institute publication policies.

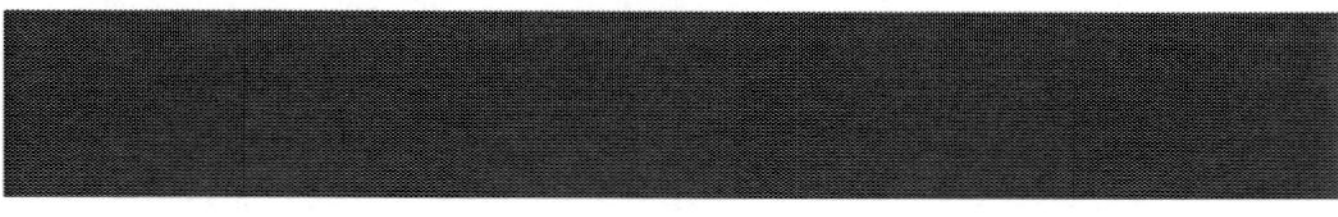

ACI member **Philipp Grosser** is a Research Engineer and Doctoral Candidate in the Department of Fastening and Strengthening Methods at the Institute of Construction Materials, University of Stuttgart, Germany. He received his diploma degree in structural engineering from the University of Karlsruhe. He is a member of the *fib* Special Activity Group, "Fastenings to Concrete and Masonry."

ACI member **Werner Fuchs** is the Director of Fastening Technology Research at the University of Stuttgart. He received his diploma degree in structural engineering from the University of Karlsruhe and his PhD from the University of Stuttgart. He is a member of ACI Committees 349, Concrete Nuclear Structures; 355, Anchorage to Concrete; and International Partnerships and Publications. He is also a member of Joint ACI-ACSE Committee 408, Development and Splicing of Deformed Bars, as well as a variety of European committees responsible for the development of code provisions in the field of fastening technology.

Rolf Eligehausen, FACI, is a retired Professor of Fastening Technology at the University of Stuttgart. He is a member of numerous national and international committees in the field of fastening technique and reinforced concrete, including ACI Committees 349, Concrete Nuclear Structures; 355, Anchorage to Concrete; and Joint ACI-ACSE Committee 408, Development and Splicing of Deformed Bars. Additionally, he is Chairman of the *fib* Special Activity Group, "Fastenings to Concrete and Masonry," and the CEN Working Group, "Design of Fastenings for Use in Concrete." He has authored or co-authored more than 300 papers and two textbooks.

APPENDIX B

Cited references are at the end of each chapter. Additional resources are noted below:

2011 North American Product Technical Guide, Volume 2: Anchor Fastening Technical Guide, Hilti.

"Adhesive Anchors In Concrete Under Sustained Tensile Loads," *Concrete International*, V. 29, No. 12, Dec. 2007, pp. 30-31.

Vuletic, Ryan, "Torque-Controlled Adhesive Anchors: An Introduction to a New Type of Post-Installed Anchor," *Structure Magazine*, May 2008, pp. 22-24.

Mattis, Lee, "CAMA Special Inspection Guidelines for Expansion and Adhesive Anchors," Copyright @ 2002 CAMA, pp. 1-7

National Transportation Safety Board. "Ceiling Collapse in the Interstate 90 Connector Tunnel, Boston, Massachusetts, July 10, 2006," Highway Accident Report NTSB/HAR-07/02, , Washington, DC, July 2007, 120 pp.

Cook, R. A.; Douglas, E.P.; and Davis, T. M, NCHRP Report 639, "Adhesive Anchors in Concrete Under Sustained Loading Conditions," Transportation Research Board

Cook, R. A., and Konz, R. C. , "Factors Influencing Bond Strength of Adhesive Anchors," *ACI Structural Journal*, V. 98, No. 1, Jan. 2001, pp. 76-86.

Eligehausen, R.; Cook, R.; and Appl, J., "Behavior and Design of Adhesive Bonded Anchors," *ACI Structural Journal*, V. 103, No. 6, Nov. 2006, pp. 822-831.

Eligehausen, R. and Silva, J., "The Assessment and Design of Adhesive Anchors in Concrete for Sustained Loading." Hilti , January 2008, pp. 1-20..

Eligehausen, R.; Hofacker, I.; and Lettow, S., "Fastening Technique – Current Status and Future Trends." Institute of Construction Materials, University of Stuttgart, Germany, Oct. 2009, pp. 1-17.

Ezrin, Myer and Storrs, CT, "Boston's Big Dig Fatal Epoxy Adhesive Failure, "Presented at Society of Plastics Engineers Annual Technical Conference (SPE ANTEC), FAPSIG, May 2008, Milwaukee, pp. 1-5.

Grosser, P.; Fuchs, W.; and Eligehausen, R, "A Field Study of Adhesive Anchor Installations: Theory and practice," *Concrete International*, V. 33, No. 1, January 2011, pp. 57-63.

Parsons Brinckerhoff, "Guide for Design and Installation of Anchor Bolts Used in Transportation Structures," Dec. 2008, Revision 0, pp. 1-82.

Handbook of Würth Dowel and Anchor Technology: Basic Principles – Applications – Practice. Adolf Würth GmbH & Co. KG. ISBN 3-89929-066-6, 2005, pp. 1-272.

Nasvik, J, "All About Anchors :An introduction to post-installed bonded and mechanical anchoring systems for concrete," Concrete Construction, Publication #C00H046, , Aug. 2000, pp. 1-3

Wollmershauser, R. E., and Mattis, Lee, "Understanding Adhesive Anchor Installation and Inspection." Concrete Anchor Manufacturer's Association (CAMA), pp. 1-11..

Powers Fasteners, "Post-Installed Anchor Technology," Technical Manual for the Design Professional 6th Edition, pp. 1-33.

Simpson Strong-Tie Anchor Systems, "Anchor Update Newsletter," Vol. 1, No. 1, Fourth Quarter 2000, pp. 1-5

Wollmershauser, R., and Mattis, L., "Adhesive Anchor Installation and Inspection," Concrete International, V. 30, No. 12, Dec. 2008, pp. 36-40.

Vuletic, R. and Pearson, J, "Use of Adhesive Anchors to Resist Long-Term Loads," Structural Engineer, Jan. 2008.

ACI Committee 318, "Building Code Requirements for Structural Concrete (ACI 318-11) and Commentary," American Concrete Institute, Farmington Hills, MI, 2011, 509 pp.

ACI Committee 355, "Qualification of Post-Installed Adhesive Anchors in Concrete (ACI 355.4-11) and Commentary," American Concrete Institute, Farmington Hills, MI, 2011, 59 pp.

AC308, "Acceptance Criteria for Post-Installed Adhesive Anchors in Concrete Elements," ICC Evaluation Service, Inc., Feb. 2012, 121 pp.

APPENDIX C

Adhesive Anchor Installer Companion Videos

Follow this QR code to view the ACI Adhesive Anchor Installer companion videos (formerly supplied as a DVD with this workbook). These videos provide an introduction to the ACI Adhesive Anchor Installer program as well as a demonstration of the overhead and vertical-down adhesive anchor installations.